AF598044

Methods in Molecular Biology

Series Editor
John M. Walker
School of Life and Medical Sciences
University of Hertfordshire
Hatfield, Hertfordshire, UK

For further volumes:
http://www.springer.com/series/7651

For over 35 years, biological scientists have come to rely on the research protocols and methodologies in the critically acclaimed *Methods in Molecular Biology* series. The series was the first to introduce the step-by-step protocols approach that has become the standard in all biomedical protocol publishing. Each protocol is provided in readily-reproducible step-by-step fashion, opening with an introductory overview, a list of the materials and reagents needed to complete the experiment, and followed by a detailed procedure that is supported with a helpful notes section offering tips and tricks of the trade as well as troubleshooting advice. These hallmark features were introduced by series editor Dr. John Walker and constitute the key ingredient in each and every volume of the *Methods in Molecular Biology* series. Tested and trusted, comprehensive and reliable, all protocols from the series are indexed in PubMed.

Fasciola hepatica

Methods and Protocols

Edited by

Martin Cancela

Laboratório de Biologia Molecular de Cestódeos, Centro de Biotecnologia, Universidade Federal do Rio Grande do Sul (UFRGS), Porto Alegre, Rio Grande do Sul, Brazil

Gabriela Maggioli

Unidad de Biología Parasitaria, Instituto de Higiene, Facultad de Ciencias, Universidad de la República, Montevideo, Uruguay

Editors
Martin Cancela
Laboratório de Biologia Molecular de Cestódeos
Centro de Biotecnologia
Universidade Federal do Rio Grande do Sul (UFRGS)
Porto Alegre, Rio Grande do Sul, Brazil

Gabriela Maggioli
Unidad de Biología Parasitaria, Instituto de Higiene
Facultad de Ciencias
Universidad de la República
Montevideo, Uruguay

ISSN 1064-3745 ISSN 1940-6029 (electronic)
Methods in Molecular Biology
ISBN 978-1-0716-0474-8 ISBN 978-1-0716-0475-5 (eBook)
https://doi.org/10.1007/978-1-0716-0475-5

This Humana imprint is published by the registered company Springer Science+Business Media, LLC, part of Springer Nature.
The registered company address is: 1 New York Plaza, New York, NY 10004, U.S.A.

Preface

In the last 20 years, advances in molecular biology techniques have emerged, expanding our knowledge in many aspects of *Fasciola hepatica* way of life. This book comprises 17 chapters, which describe basic and advanced protocols to study *F. hepatica* parasite biology. We start this book describing protocols to obtain and maintain different developmental stages of *F. hepatica* at the laboratory for different applications such as molecular/cellular biology studies and vaccination trials (Chapter 1). Besides, Chapter 2 refers to protocols to evaluate host tissue changes induced during *F. hepatica* acute and chronic infection by analyzing peritoneal cells and hepatic lesions in sheep. Chapters 3–7 describe a diverse set of protocols to obtain and characterize tegumental and secreted proteins (Chapter 3) and extracellular vesicles (Chapter 4) by using proteomic approaches and also to perform functional studies by producing recombinant protein in *E. coli* (Chapter 5). A very detailed protocol to perform gene knockdown by using the RNAi technique in *F. hepatica* (Chapter 6) including protocol to obtain in-house double strand RNA by in vitro transcription is presented in this book. In Chapter 7, we introduce two advanced protocols to study spatial and temporal gene expression patterns using in situ and in toto hybridization techniques in different *F. hepatica* developmental stages.

The second part of this book (Chapters 8–15) is dedicated to introduce the readers to a complex repertoire of molecular biology techniques to study immunological changes induced by *F. hepatica* antigens and experimental infection. In Chapter 8, we start with a revision in molecular aspect of *F. hepatica* infection related to the evasion of host immune response. Excretion–secretion products (ESP) of *F. hepatica* are well-known to be involved in the evasion mechanisms. Chapters 9–12 describe different protocols to evaluate immunological effects of ESP on macrophages (Chapter 9), eosinophil apoptosis (Chapter 10), macrophages alternative activation, (Chapter 11), and toll-like receptor interactions (Chapter 12). Chapter 13 describes protocols to obtain peritoneal and splenic dendritic cells of mice experimentally infected by *F. hepatica* and the effect of adult worm extracts on DC maturation.

Finally, we provide protocols to perform in silico analysis to find immunogenic epitopes in *F. hepatica* protein antigens (i.e., leucine aminopeptidase and cathepsins) and the design of a chimeric protein to be used in vaccination trials (Chapter 14). A vaccination protocol in a ruminant model to evaluate the immunoprotective effect of *F. hepatica* antigens (i.e., LAP) is described (Chapter 15). Also, we include some protocols to detect *F. hepatica* albendazole resistance (Chapter 16) and for the screening of small molecules as enzyme inhibitors (Chapter 17) to identify novel antiparasitic compounds.

We selected different topics in areas such as biochemistry, immunology, molecular biology, microscopy, and vaccinology in order to extend the interest in this book to research on community working with *Fasciola hepatica* and related trematodes.

This book is the effort of many researchers who share their time and expertise, and without them it would be impossible to culminate this work. We would like to express our warm thanks to their contributions and patience.

Porto Alegre, Rio Grande do Sul, Brazil ***Martin Cancela***
Montevideo, Uruguay ***Gabriela Maggioli***

Contents

Contributors

DANIEL ACOSTA • *División Laboratorios Veterinarios "DILAVE", Unidad de Biotecnología, Ministerio de Ganaderia Agricultura y Pesca, Montevideo, Uruguay*

GUZMÁN I. ÁLVAREZ • *Laboratorio de I+D de Moléculas Bioactivas, Departamento de Ciencias Biológicas. CENUR Litoral Norte—Sede Paysandú, Universidad de la República, Paysandú, Uruguay*

LUIS I. ALVAREZ • *Laboratorio de Farmacología. Centro de Investigación Veterinaria de Tandil (CIVETAN). UNCPBA-CICPBA-CONICET, Facultad de Ciencias Veterinarias. Campus Universitario, Tandil, Argentina*

MAURICIO A. CABRERA • *Laboratorio de I+D de Moléculas Bioactivas. Departamento de Ciencias Biológicas. CENUR Litoral Norte—Sede Paysandú, Universidad de la República, Paysandú, Uruguay*

MARTIN CANCELA • *Laboratório de Biologia Molecular de Cestódeos, Centro de Biotecnologia, Universidade Federal do Rio Grande do Sul, Porto Alegre, Rio Grande do Sul, Brazil*

CANDELA CANTON • *Laboratorio de Farmacología. Centro de Investigación Veterinaria de Tandil (CIVETAN). UNCPBA-CICPBA-CONICET, Facultad de Ciencias Veterinarias. Campus Universitario, Tandil, Argentina*

CARLOS CARMONA • *Unidad de Biología Parasitaria. Instituto de Higiene. Facultad de Ciencias, Universidad de la República, Montevideo, Uruguay*

ESTELA CASTILLO • *Sección Bioquímica. Facultad de Ciencias, Universidad de la Republica, UdelaR, Montevideo, Uruguay*

LAURA CEBALLOS • *Laboratorio de Farmacología. Centro de Investigación Veterinaria de Tandil (CIVETAN). UNCPBA-CICPBA-CONICET, Facultad de Ciencias Veterinarias. Campus Universitario, Tandil, Argentina*

LAURA CERVI • *Laboratorio de Parasitología y Micología, Departamento de Bioquímica Clínica, Facultad de Ciencias Químicas, Universidad Nacional de Córdoba. Haya de la Torre y Medina Allende, Ciudad Universitaria, Córdoba, Argentina; Centro de Investigaciones en Bioquímica Clínica e Inmunología (CIBICI), CONICET. Haya de la Torre y Medina Allende, Ciudad Universitaria, Córdoba, Argentina*

LAURA S. CHIAPELLO • *Laboratorio de Parasitología y Micología, Departamento de Bioquímica Clínica, Facultad de Ciencias Químicas, Universidad Nacional de Córdoba. Haya de la Torre y Medina Allende, Ciudad Universitaria, Córdoba, Argentina; Centro de Investigaciones en Bioquímica Clínica e Inmunología (CIBICI), CONICET. Haya de la Torre y Medina Allende, Ciudad Universitaria, Córdoba, Argentina*

ILEANA CORVO • *Laboratorio de I+D de Moléculas Bioactivas. Departamento de Ciencias Biológicas. CENUR Litoral Norte—Sede Paysandú, Universidad de la República, Paysandú, Uruguay*

EDUARDO DE LA TORRE-ESCUDERO • *Institute for Global Food Security. School of Biological Sciences, Queen's University Belfast, Belfast, UK*

NICOLÁS DELL'OCA • *Departamento de Genética. Facultad de Medicina, Universidad de la República, UdelaR, Montevideo, Uruguay*

A. ESCAMILLA • *Department of Anatomy and Comparative Pathology. Faculty of Veterinary Medicine, University of Córdoba, Córdoba, Spain*

Florencia Ferraro • *Laboratorio de I+D de Moléculas Bioactivas. Departamento de Ciencias Biológicas. CENUR Litoral Norte—Sede Paysandú, Universidad de la República, Paysandú, Uruguay*

Olgary Figueroa-Santiago • *Department of Biology, University of Puerto Rico at Cayey, Cayey Puerto Rico, USA*

Robin J. Flynn • *Department of Infection Biology, Institute of Infection and Global Health, University of Liverpool, Liverpool, UK*

Teresa Freire • *Grupo de Inmunomodulación y Desarrollo de Vacunas, Departamento de Inmunobiología, Facultad de Medicina, Universidad de La República, Montevideo, Uruguay*

Federico Fossa • *Unidad de Biología Parasitaria. Instituto de Higiene. Facultad de Ciencias, Universidad de la República, Montevideo, Uruguay*

Alicia Galiano • *Àrea de Parasitologia, Departament de Farmàcia i Tecnologia Farmacèutica i Parasitologia, Universitat de València, Burjassot, Valencia, Spain*

Valeria Gayo • *División Laboratorios Veterinarios "DILAVE", Unidad de Biotecnologia, Ministerio de Ganaderia Agricultura y Pesca, Montevideo, Uruguay*

Lorena Guasconi • *Laboratorio de Parasitología y Micología, Departamento de Bioquímica Clínica, Facultad de Ciencias Químicas, Universidad Nacional de Córdoba. Haya de la Torre y Medina Allende, Ciudad Universitaria, Córdoba, Argentina; Centro de Investigaciones en Bioquímica Clínica e Inmunología (CIBICI), CONICET. Haya de la Torre y Medina Allende, Ciudad Universitaria, Córdoba, Argentina*

Uriel Koziol • *Sección Biología Celular. Facultad de Ciencias, Universidad de la República, Montevideo, Uruguay*

Carlos E. Lanusse • *Laboratorio de Farmacología, Centro de Investigación Veterinaria de Tandil (CIVETAN), UNCPBA-CICPBA-CONICET, Facultad de Ciencias Veterinarias, Campus Universitario, Tandil, Argentina*

Gabriela Maggioli • *Unidad de Biología Parasitaria. Instituto de Higiene. Facultad de Ciencias, Universidad de la República, Montevideo, Uruguay*

Antonio Marcilla • *Àrea de Parasitologia, Departament de Farmàcia i Tecnologia Farmacèutica i Parasitologia, Universitat de València, Burjassot, Valencia, Spain; Joint Research Unit on Endocrinology, Nutrition and Clinical Dietetics, Health Research Institute La Fe, Universitat de Valencia, Burjassot, Valencia, Spain*

A. Martínez-Moreno • *Department of Animal Health (Parasitology), Faculty of Veterinary Medicine, University of Córdoba, Córdoba, Spain*

F. J. Martínez-Moreno • *Department of Animal Health (Parasitology), Faculty of Veterinary Medicine, University of Córdoba, Córdoba, Spain*

María Martinez Valladares • *Departamento de Salud Animal, Instituto de Ganadería de Montaña (CSIC-Universidad de León), Grulleros, Spain*

Diana T. Masih • *Laboratorio de Parasitología y Micología, Departamento de Bioquímica Clínica, Facultad de Ciencias Químicas, Universidad Nacional de Córdoba. Haya de la Torre y Medina Allende, Ciudad Universitaria, Córdoba, Argentina*

Maria Teresa Minguez • *Secció de Microscopia, Serveis Centrals de Suport a l'Investigació Experimental, Universitat de València, Burjassot, Valencia, Spain*

V. Molina-Hernández • *Department of Animal Health (Parasitology), Faculty of Veterinary Medicine, University of Córdoba, Córdoba, Spain*

Mayowa Musah-Eroje • *School of Veterinary Medicine and Science, University of Nottingham, Nottingham, UK*

I. L. Pacheco • *Department of Anatomy and Comparative Pathology, Faculty of Veterinary Medicine, University of Córdoba, Córdoba, Spain*

J. Pérez • *Department of Anatomy and Comparative Pathology, Faculty of Veterinary Medicine, University of Córdoba, Córdoba, Spain*

R. Pérez • *Department of Animal Health (Parasitology), Faculty of Veterinary Medicine, University of Córdoba, Córdoba, Spain*

Gabriel Rinaldi • *Wellcome Sanger Institute, Wellcome Genome Campus, Hinxton, Cambridgeshire, UK*

Mark W. Robinson • *Institute for Global Food Security, School of Biological Sciences, Queen's University Belfast, Belfast, UK*

M. T. Ruiz-Campillo • *Department of Anatomy and Comparative Pathology, Faculty of Veterinary Medicine, University of Córdoba, Córdoba, Spain*

Cecilia Salazar • *Unidad de Biología Parasitaria, Instituto de Higiene, Facultad de Ciencias, Universidad de la República, Montevideo, Uruguay; Microbial Genomics Laboratory, Institut Pasteur de Montevideo, Montevideo, Uruguay*

Christian M. Sánchez • *Àrea de Parasitologia, Departament de Farmàcia i Tecnologia Farmacèutica i Parasitologia, Universitat de València, Burjassot, Valencia, Spain; Joint Research Unit on Endocrinology, Nutrition and Clinical Dietetics, Health Research Institute La Fe, Universitat de Valencia, Burjassot, Valencia, Spain*

Marianela C. Serradell • *Laboratorio de Parasitología y Micología, Departamento de Bioquímica Clínica, Facultad de Ciencias Químicas, Universidad Nacional de Córdoba. Haya de la Torre y Medina Allende, Ciudad Universitaria, Córdoba, Argentina; Centro de Investigación y Desarrollo en Inmunología y Enfermedades Infecciosas (CIDIE), Consejo Nacional de Investigaciones Científicas y Técnicas (CONICET)/Universidad Católica de Córdoba (UCC), Córdoba, Argentina*

José F. Tort • *Departamento de Genética, Facultad de Medicina, Universidad de la República, UdelaR, Montevideo, Uruguay*

R. Zafra • *Department of Animal Health (Parasitology), Faculty of Veterinary Medicine, University of Córdoba, Córdoba, Spain*

Chapter 1

Maintenance of Life Cycle Stages of *Fasciola hepatica* in the Laboratory

Valeria Gayo, Martin Cancela, and Daniel Acosta

Abstract

Fasciola hepatica has a heteroxenous complex life cycle that alternates between an invertebrate intermediate and a mammalian definitive host. The life cycle has five well-defined phases within their hosts and the environment: (1) eggs released from the vertebrate host to the environment and its subsequent development; (2) emergence of miracidia and their search and penetration into an intermediate snail host; (3) development and multiplication of larval stages within the snail; (4) emergence of cercariae and the encystment in metacercariae; and (5) ingestion of infective metacercariae by the definitive host and development to its adult form. Here we describe some protocols to obtain and maintain different developmental stages of *F. hepatica* in the laboratory for different applications (molecular/cellular biology studies, vaccination trials, etc.).

Key words Life cycle, Developmental stages, Definitive host, Intermediate host, Snail

1 Introduction

Fasciolosis caused by *F. hepatica* is widely recognized as one of the most important helminthiasis affecting cattle and sheep in terms of geographical distribution and economic impact on farmers. Its relevance as a zoonotic agent in regions of Latin America [1] and Africa [2] is also emerging with millions of people at risk of infection [3, 4].

Taking into account the broad distribution of fasciolosis and the complexity of *F. hepatica* life cycle, there is an increased interest to improve the ways of maintaining the cycle under controlled laboratory conditions. This would allow for a better understanding about the biology of the snail, the intermediate-host and, at the same time, obtaining the vertebrate host infective stage [5]. The intermediate host, a snail, is highly specific and is where parthenogenetic larval multiplication takes place. The vertebrate definitive host, in which sexual reproduction occurs, has a low specificity and can be a wide range of mammals, including human, and also birds

Martin Cancela and Gabriela Maggioli (eds.), *Fasciola hepatica: Methods and Protocols*, Methods in Molecular Biology, vol. 2137, https://doi.org/10.1007/978-1-0716-0475-5_1,

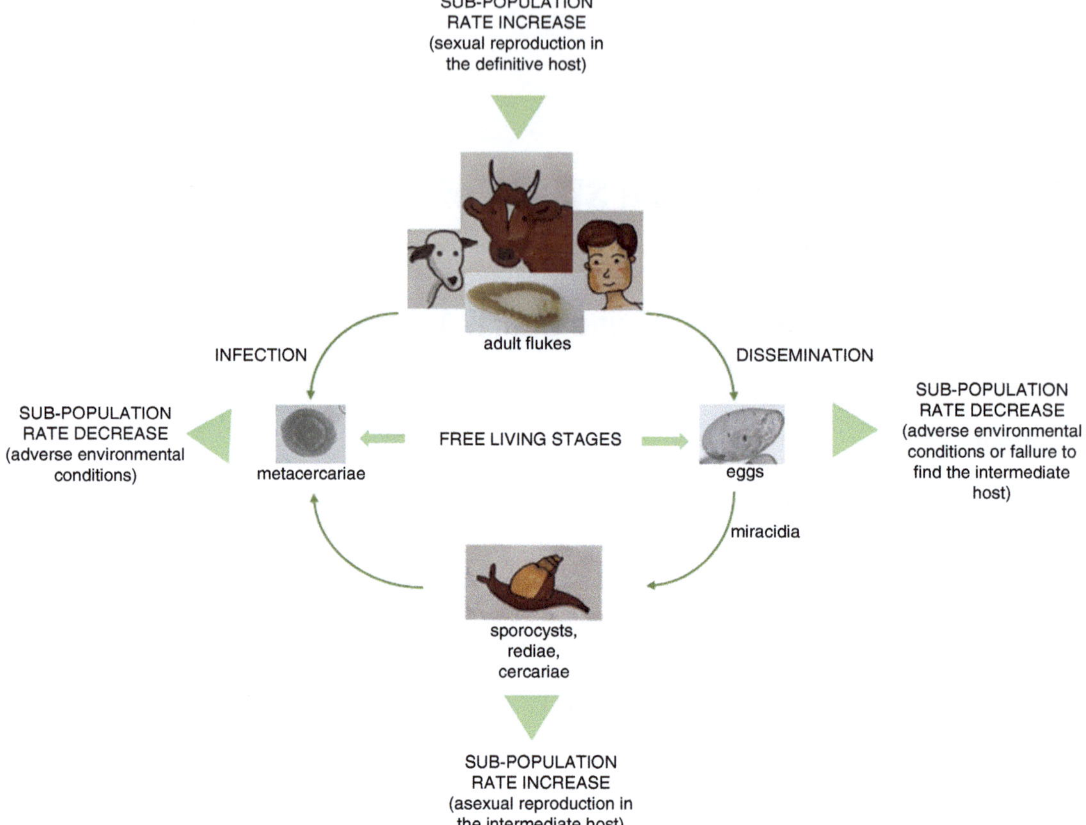

Fig. 1 *Fasciola hepatica* life cycle

of the ratite family (Fig. 1). Thus, the obtained metacercariae can be used to carry out in vivo and in vitro studies for different purposes. One of the objectives of this chapter is to introduce the readers to maintenance of the intermediate snail host and production of metacercaria, the infective larval stage for the mammalian host.

To ensure this, we describe protocols for maintenance of lymnaeids in the laboratory, obtention of *Fasciola hepatica* eggs and miracidia, infection of the intermediate host with miracidia, cercariae emission, and encystment in metacercariae. Also, we describe an in vitro metacercaria excystment protocol to obtain newly excysted juveniles of *F. hepatica*.

2 Materials

2.1 Maintenance of Lymnaea neotropica Colony in the Laboratory

2.1.1 Snail Culture

1. Petri dishes.
2. Metal or plastic trays.
3. Incubator.
4. Refrigerator.

2.1.2 Oscillatoria Culture

1. 75% sterilized clay soil naturally obtained from the field.
2. 25% sterilized organic soil naturally obtained from the field.
3. Calcium carbonate analytical grade.
4. Incubators 20–22 °C and 24–26 °C.
5. 400 W lamp.

2.2 Obtaining Eggs from F. hepatica

1. Gall bladder from infected cattle.
2. Distilled water.
3. Sieves with different mesh size 30–500 μm.
4. Plastic tubes: 50, 15, and 1.5 mL.
5. Aluminum foil.
6. Refrigerator (4 °C).

2.3 Hatching of Miracidia by Incubation of Eggs

1. Incubator (25 °C).
2. Plates dishes.
3. *F. hepatica* eggs.
4. Distilled water.
5. Aluminum foil.
6. Stereomicroscope.

2.4 Snail Infection

1. *L. neotropica* snail 4.0–5.0 mm length.
2. Miracidia.
3. Glass tubes.
4. Distilled water.

2.5 Development of F. hepatica Inside the Snail and Metacercariae Obtention

1. Scissors.
2. Tweezers.
3. Cellophane bags.

2.6 In Vitro Excystment of F. hepatica Metacercariae

All materials used for in vitro excystment should be sterile and suitable for cell culture.

All materials in contact with live metacercariae have to be disinfected using 10% sodium hypochlorite.

2.6.1 Plasticware

1. Polystyrene sterile 6-well culture plates.
2. Petri dishes.
3. Plastic tubes: 15 and 50 mL.
4. Pasteur pipette.
5. 0.22 μm syringe filters.
6. 6- or 12-well Netwell inserts (74 μm and 20 μm mesh strainers).
7. Syringe: 20–50 mL volume.
8. 10-kDa membrane concentration tubes (15 mL).

2.6.2 Solutions

1. Sterile PBS: Dissolve in 800 mL distilled water (dH_2O), 2.56 g $Na_2HPO_4{\cdot}7H_2O$, 8 g NaCl, 0.2 g KCl, and 0.2 g KH_2PO_4. Adjust the pH to 7.3. Add dH_2O to the total volume of 1 L and autoclave.
2. Antibiotic: gentamycin or penicillin solution (10,000 U/mL).
3. Antifungal: amphotericin B solution (250 μg/mL).
4. 1% sodium hypochlorite to be freshly prepared.
5. 50 mM HCl solution: add 210 μL of 12 M HCl to 50 mL of dH_2O.
6. 1% $NaHCO_3$/0.8% NaCl pH 8.0: dissolve 0.5 g of $NaHCO_3$, 0.4 g NaCl in 40 mL of dH_2O, adjust pH to 8.0 and volume to 50 mL.
7. 3.3 M L-cysteine (100× stock): dissolve 0.578 g of L-cysteine in 1 mL H_2O. Stock in 50 μL aliquot at −20 °C (thaw once and discard).
8. 10% sodium taurocholate: dissolve 10 g of sodium taurocholate in 100 mL of dH_2O store at −20 °C (thaw once and discard).
9. Solution A: 50 mM HCl and 33 mM L-cysteine.
10. Solution B: 1% $NaHCO_3$/0.8% NaCl pH 8.0. Just before use add the sodium taurocholate to a 0.2% v/v concentration.
11. RPMI-1640 medium: dissolve 10.4 g RPMI-1640 with L-glutamine in 800 mL of MilliQ water and add 2 g sodium bicarbonate. Supplement with gentamicin (100 U/mL) and amphotericin B (2 μg/mL). Adjust pH to 7.3 and final volume to 1 L. Sterilize by filtration using a 0.22 μm membrane in a laminar flux cabin.

12. Silicon.
13. PBS-5% fetal bovine serum (FBS).
14. Micro BCA assay kit.

2.6.3 Equipment and Instruments

1. Water bath.
2. CO_2 incubator.
3. Stereoscopic microscope.
4. Scalpel.
5. Scissors.
6. Tweezers.

2.7 *Obtaining* F. hepatica *Adult Flukes*

1. 500 mL flasks.
2. Knife.
3. Gloves.
4. Water bath at 37 °C.
5. PBS.
6. Plastic trays.
7. Spoons.
8. Modified RPMI-1640 medium: dissolve 10.4 g RPMI-1640 with L-glutamine in 800 mL of MilliQ water and add 7.15 g de HEPES and 20 g glucose. Supplement with gentamicin (100 U/mL). Adjust pH to 7.3 and final volume to 1 L. Sterilize by filtration using a 0.22 μm membrane in a laminar flow cabinet.

3 Methods

3.1 *Maintenance of* Lymnaea neotropica *Colony in the Laboratory*

The methods of culture consist, essentially on simulating the natural habitat of the snail in earthenware pans, metal or plastic trays or in shallow glass dishes containing a layer of mud. Different sources of food have been tested to maintain and achieve reproduction of the snails in the laboratory: *Oscillatoria* and lettuce.

Oscillatoria is a cyanobacterium living in fresh water that owes its name to its oscillatory movement. This cyanobacterium has photosynthetic capacity and has proved to be a very satisfactory food for snails [6].

3.1.1 Oscillatoria *Culture*

1. Prepare glass dishes by mixing 75% sterilized clay soil, 25% sterilized organic soil, calcium carbonate over 0.3 mekv/L in distilled water until a muddy paste is obtained. This paste will serve as a support.

2. Seed small pieces of *Oscillatoria* collected from watercourses at several points on the plate (or metal/plastic trays or in shallow glass dishes) containing a layer of mud. Incubate overnight under a 400 W lamp (*see* **Note 1**).
3. After the overnight incubation, the *Oscillatoria* is spread gently over the entire surface and left under the lamp at 20 °C until the whole area is covered (approximately 48 h). The dishes can be stored at 4 °C for at least 1 week until they must be used for food.

3.1.2 Lymnaea neotropica *Colony*

In 1973, the *Departamento de Parasitologia* of the current *División Laboratorios Veterinarios "DILAVE", Ministerio de Ganaderia Agricultura y Pesca, Montevideo, Uruguay* began to study the life cycle of *F. hepatica*, by collecting from the field and experimentally infecting a probable variety of snail hosts. Later through the FAO Project URU 78/008, Dr. C.B Ollerenshaw from the Central Veterinary Laboratories, Weybridge, England came to Uruguay and offered advice in relation to the maintenance of the cycle under laboratory conditions and collecting from the field both the actual strain of snail host and the *Oscillatoria* cyanobacteria to feed them.

The snail colony maintained at our lab was first classified as *Lymnaea viatrix* [7], but latter molecular tools forced to rename the species. Sequence analysis of the ITS-1 amplicons showed 99.6% identity with the previously reported sequence of *Lymnaea neotropica* (GenBank, accession number AM412228).

1. To propagate snail colony use 10 cm diameter petri dishes with 2 snails or 20 snails in 30 cm petri dishes (*see* **Note 2**).
2. Collect the egg masses laid by the snails that contain a variable number of eggs (ranging from 8 to 13) (*see* **Note 3**).
3. Place egg masses in distilled water at 20–22 °C. Hatching begins 6 days later.
4. Feed newborn snails with *Oscillatoria* algae and maintain at 20–22 °C under high humidity conditions (*see* **Note 4**).

3.2 Obtaining Eggs from F. hepatica

1. Collect the gall bladder content obtained from *F. hepatica* infected cattle at an abattoir.
2. Wash the eggs several times with distilled water until a transparent solution is obtained. When possible the use of a set of sieves with different mesh size result in a faster procedure (*see* **Note 5**).
3. Resuspend the eggs in distilled water and store at 4 °C protected from light (*see* Fig. 2a and **Note 6**).

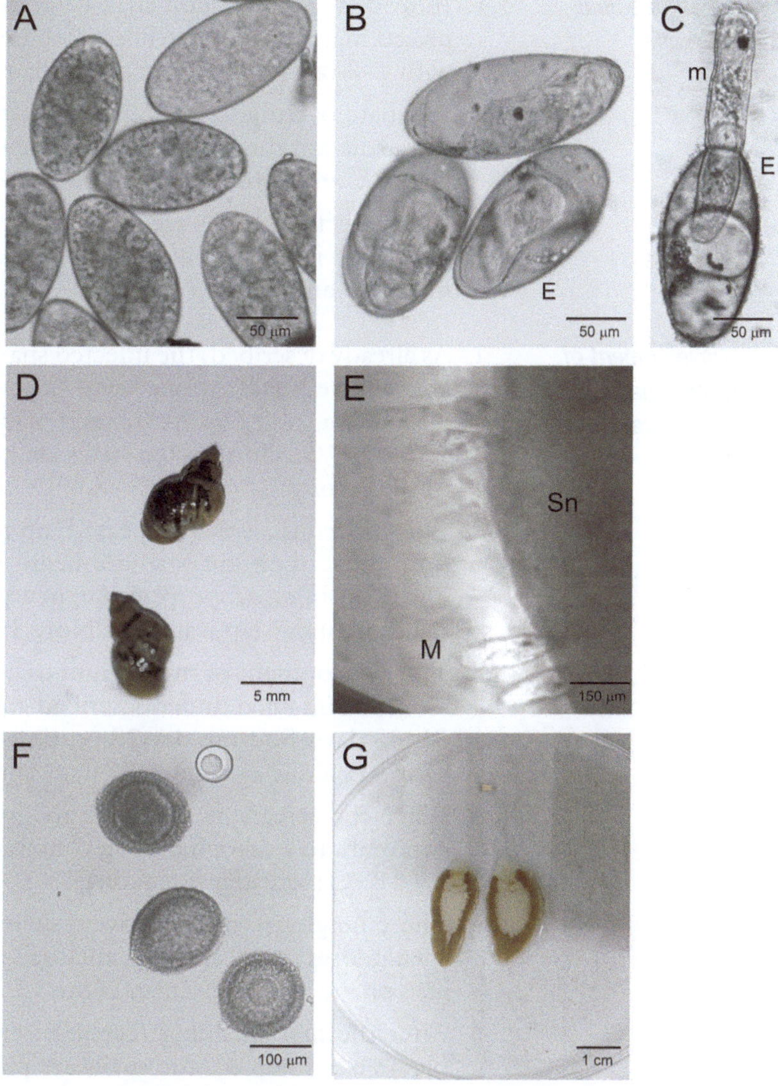

Fig. 2 *F. hepatica* developmental stages. (**a**) Embryonated eggs (E) collected from the gallbladder, (**b**) embryonated eggs after 10 days at 25 °C, (**c**) miracidia hatching (m) after 14 days at 25 °C, (**d**) *L. neotropica* snail, (**e**) snail (Sn) infection by miracidia (M), (**f**) Metacercaria and (**g**) adult flukes

3.3 Hatching of Miracidia by Incubation of Eggs

1. Incubate 20 μL of egg pellet in 20 mL of distilled water at 25–26 °C (Fig. 2a, *see* **Note 7**).
2. Monitor embryo development at different times (2 days intervals) by using a microscope or stereomicroscope. Miracidium can be observed after 10–12 days inside the egg (*see* Fig. 2b).
3. Expose to light to induce miracidia hatching (Fig. 2c, *see* **Note 8**).

3.4 Snail Infection

1. Individual *L. neotropica* snail (Fig. 2d) 4.0–5.0 mm length is placed in a 5.0 mL tube filled with distilled water and infected with 1–3 recently hatched miracidia (*see* Fig. 2c, **Note 9**).
2. Incubate overnight at room temperature under artificial light. Miracidia attached to snail can be observed during this period (Fig. 2e, *see* **Note 10**).
3. After infection, snails are placed into petri dishes with *Oscillatoria* and incubated at 24–26 °C (incubator) with constant humidity and light.

3.5 Development of F. hepatica Inside the Snail and Metacercaria Obtention

1. As already described, the infected snail maintained at incubator are removed daily, washed with tap water, transferred to new plates with fresh *Oscillatoria* and placed again at the incubator at 24–26 °C. The infected snails are kept isolated from the rest of the colony.
2. This procedure is daily repeated, up to 25 days after the infection. Since then the snail are begin to be examined carefully under the microscope, placed in new plates but without placing the snail under tap water (*see* **Note 11**).
3. The development of miracidium to cercaria inside the infected snails, maintained under described conditions, demands about 31–32 days. So from day 28 postinfection one snail is crushed daily to see the evolutionary state of redia to cercaria.
4. When the cercaria is observed, the infected snail is transferred overnight to a cellophane bag containing fresh distilled water at 15–18 °C (cercariae emission).
5. The emerging cercaria actively swims for a couple of hours, then it loses its tail and gets enclosed into a cyst attached to the cellophane bag (Fig. 2f, *see* **Note 12**).
6. After the first shedding (cercariae emission) inside the cellophane bag (**step 4**), the snails are placed again in petri dishes with Oscillatoria at 24–26 °C for a week. This is because not all the stages of development within the snails reach the cercarial stage at the same time.
7. After 1 week, repeat **step 4**.
8. After the second emission, infected snails are placed again in petri dishes at 24–26 °C, and this process can be repeated so many times until the entire population of metacercariae is obtained.

3.6 In Vitro Metacercariae Excystment

1. In a petri dish count encysted metacercariae attached to cellophane bags (*see* **step 5** Subheading 3.5) (*see* **Note 13**).
2. Carefully detach encysted metacercariae from cellophane using a scalpel.

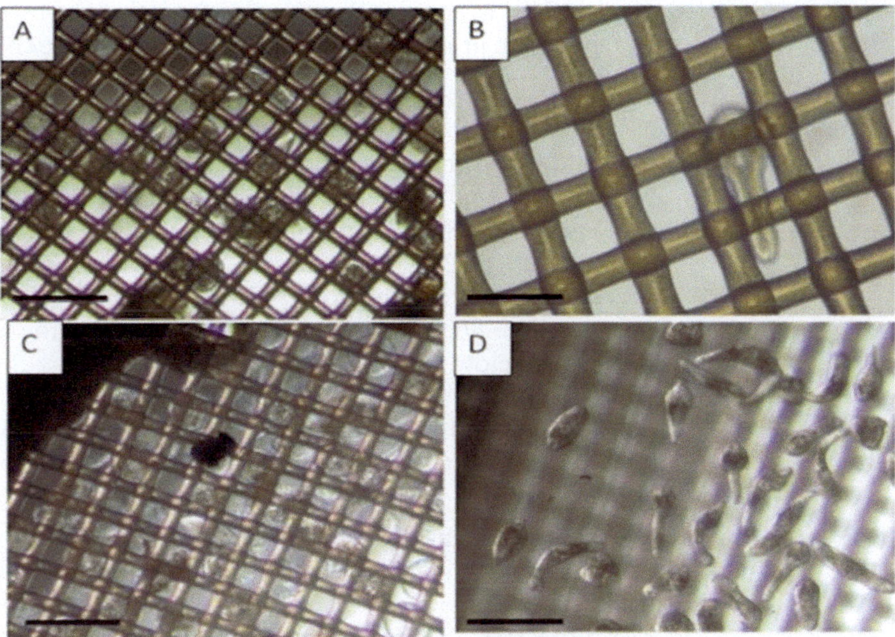

Fig. 3 In vitro excystment of metacercariae. (**a**) Metacercariae (MC) over the 70 μm net, with some NEJ emerging, (**b**) newly excysted juveniles (NEJs) that migrated to the bottom of the well. (**c**) Mostly empty MC cysts are retained on the mesh while and (**d**) NEJs concentrate in the bottom of the well. Scale Bar: 200 μm in (**a**, **c**, and **d**, and 80 μm in **b**)

3. Use a drop of water to collect metacercariae inside a 74 μm filter (Fig. 3a, *see* **Note 14**).
4. In a 6-well plate put the 74 μm filter and add 4 mL of 1% sodium hypochlorite for 5 min at 39 °C to remove external cyst wall. See under stereoscopic microscope if external wall has been efficiently removed. This step is critical in order to achieve a high efficiency in metacercaria excystation.
5. Wash ten times with PBS at 39 °C.
6. Mix equal volume of solution A and B (AB solution). After mixing A and B solutions (AB solution), CO_2 production (air bubbles) has to be observed (*see* **Note 15**).
7. Sterilize AB solution by filtration using a 0.22 μm syringe filter.
8. Put the 74 μm filter unit with metacercariae into a clean well and incubate in AB solution for 90–180 min at 39 °C. Newly excysted juveniles will begin to emerge from the inner capsules in the first hour of incubation. Track metacercaria excystment by using a stereoscopy microscope every 15 min (Fig. 3b–d, *see* **Note 16**).
9. Metacercaria excystment will demand approximately 2 or 3 h (*see* **Note 17**).

10. Collect NEJ as soon as they emerge from the cyst by pipetting and put them in a 20 μm filter unit or 1.5 mL tube (*see* **Note 18**).
11. Wash NEJ by sedimentation using RPMI-1640 medium or PBS at 37 °C to eliminate any traces of excystation medium.
12. Depending on the technique to be used, we can preserve NEJ in different conditions. For protein analysis, you can freeze the NEJ pellet at −80 °C. For microscopy, fix NEJ in 4% paraformaldehyde in PBS (*see* Chapter 7; Subheading 3.4). For nuclear acid extraction, store NEJ pellet in 500 μL of TRIzol reagent at −80 °C. For NEJ cultivation, see the protocol described below.

3.7 NEJ Culture

1. Collect NEJ inside 20 μm filter and incubate in RPMI-1640 medium (1 NEJ/μL medium) at 37 °C with 5% CO_2 atmosphere.
2. For proteomic analysis, change RPMI-1640 medium without fetal bovine serum (FBS) daily and collect the culture supernatant that contains the NEJ excretion–secretion products (ESP) (*see* **Note 19**).
3. Centrifuge at 20,000 × *g* for 30 min at 4 °C.
4. Filter NEJ ESP using 0.22 μm membrane and store at −20 °C for further analysis.
5. For mass spectrometry analysis, NEJ ESP can be 10–20 fold concentrated using a 3-kDa cutoff membrane concentration tubes.
6. Determine protein concentration using micro BCA assay kit.

3.8 Obtaining F. hepatica Adult Flukes

F. hepatica adult flukes can be obtained from different hosts: natural (e.g., ovine, bovine) and experimental infected (e.g., mouse, mice, rabbit, ovine, and bovine) during vaccination trials (*see* Chapter 15). We describe a protocol to obtain and transport adult fluke collected from bovine livers obtained from the local abattoir.

1. Inspect bovine liver to detect fibrotic and hypertrophied bile duct.
2. Cut transversally the bile duct with the help of a knife.
3. In a 500 mL flask, collect the bile from the gallbladder of the same bovine liver that you are collecting the parasite (*see* **Note 20**).
4. Maintain the bile at 37 °C in a water bath (*see* **Note 21**).
5. Squeeze the bile duct to push out the flukes.
6. Put the adult flukes collected inside a 500 mL flask prefilled with fresh bile.

7. Repeat this procedure until all fibrotic and hypertrophied bile ducts have been examined.
8. Transport parasites to the lab in the bile at 37 °C.
9. Once at the lab, parasites are exhaustively washed with PBS or RPMI until *F. hepatica* gut content has been regurgitated (*see* **Note 22**).
10. Live *F. hepatica* can be used in downstream protocols including culture in different conditions (modified RPMI-1640 or other culture media) to obtain excretion–secretion products (*see* Chapter 12, Subheading 3.1) or frozen in N_2 (l) to prepare protein extracts. For nucleic acid extraction, store one adult fluke (0.1 g) in 1 mL TRIzol reagent (*see* Chapter 5, Subheading 3.1). For histological analysis, adult fluke can be fixed in 4% paraformaldehyde in PBS (Fig. 2g, *see* Chapter 7).

4 Notes

1. *Oscillatoria* should be slightly hydrated and placed spaced 5 cm.
2. *Lymnaeids* are dextrogyrous air-breathing freshwater pulmonate gastropod snails. They inhabit both temporary and permanent freshwater bodies and can be found in rather alkaline shallow well-aerated ponds, lakes, streams, rivers, water ditches, irrigation channels, and so on. Under hot and dry conditions snails are capable of surviving for very extended periods in a state of drought-induced dormancy lasting from 6 weeks to a year [8]. They are hermaphrodite, and most of the genera belonging to the Lymnaeidae family are facultatively self-fertile.
3. These snails begin to lay eggs between the fourth and the sixth week after hatching and continue to do so daily until approximately week 16 when the snail has reached a length of 1 cm. We have observed that a snail immediately isolated after hatching lays large numbers of fertile eggs. We have registered that during the laying period that an individual *L. neotropica* could produce between 600 and 800 new eggs [7]. Our observations suggested that larger snails produced more eggs and that the number increased with the age and size of the snail.
4. Considering that in nature, snails feed on degraded plants, boiled lettuce is also used as a natural food at laboratories. Alternatively, artificial source of food for fishes or wheat germ could be used depending on the size of the breeding colony and if the aim is to develop experimental studies or to obtain large quantity of *F. hepatica* living stages. For a successful culture, it is necessary to have an adequate exposure to light,

because it has been proved that plants in a state of active photosynthesis stimulated the activity of the snails and promoted their growth and reproduction.

5. The bile is passed through two coupled sieves: the upper one of 200 μm sieve to retain gross debris and the lower one of 35 μm sieve to retain clean *F. hepatica* eggs.
6. The maintenance, incubation, and hatching of *F. hepatica* eggs is controlled by various factors, the most important being water, but also temperature, light, salinity and others like dissolved oxygen, pH, and ionic charge of the water. For these reasons, after washing gently with a jet of water the eggs are resuspended in distilled deionized water and placed in a brown jar at 4 °C for a maximum period of a year until used.
7. When placed at temperatures above 10 °C, the eggs develop slowly to miracidia which have an ovoid, ciliated, and elongated shape, about 130–180 μm long.
   As the incubation temperature increases, the maturation accelerates so that at 25–26 °C hatching occurs in a period of 12–14 days.
8. At this developmental stage, a pair of photosensitive plates (in an X shape called eye plates or ocular spots) in the anterior third part of the body is found. This is a photoreceptor organ; the emergence of miracidium from the egg operculum is stimulated when they are placed under a light due to the activation of positive phototropism and negative geotropism [9].
9. The miracidia have only a rudimentary intestine, and since they do not feed, they completely depend for their activity on endogenous food reserves. These rarely last more than 24 h, which limits the search for a host snail of this stage, after which they die. Digenetic trematodes exhibit an impressive array of sophisticated sensory and motor mechanisms well adapted to the necessity of finding, recognizing, and infecting the appropriate host as it contributes to the dissemination and perpetuation of parasites.
10. During the first 2 h, the chemotactic organic and inorganic stimulants attract the miracidium to the snail. Although the miracidium tries to penetrate the surface of the snail almost none achieves it. The miracidium finally penetrates the hepatopancreas by histolytic and mechanical action, and parthenogenetic multiplication begins to take place.
11. During a period of 3 years, we infected twenty snails monthly and exposed them to the environment. From day 30 postinfestation weekly dissection of snails showed that with temperature at 24 °C (summer) the development from miracidia to cercaria takes 39 days but when the snails were exposed to the

environment under temperature of 15 °C (autumn) it takes 100 days to complete the development to cercaria [10].

12. The encysted cercariae (called metacercaria) remain secure in their cysts until eaten by a mammalian host. If the metacercaria encystment takes place in the water and metacercariae is maintained at 4 °C, they can live for a year. However, if they encyst on grass, then they survive only for a few weeks because they can withstand short periods of drying.
13. Wear suitable gloves and isolate the working area to avoid contamination and to protect you from infection. To manipulate and cut cellophane bags, use tweezers and scissors.
14. We use commercially available 74 μm filter to facilitate the separation of metacercariae from NEJ during excystation process. NEJ passes through the 74 μm mesh, while metacercariae are retained in the filter.
15. Solutions A and B should be prepared the same day that they will be used (excystment buffer) and mixed just before use.
16. Replace excystment medium once an hour until excystation is complete. Some components of the excystation medium would precipitate after 1 h of incubation.
17. To assure a high excystment efficiency, you have to use fresh metacercariae and excystment solutions.
18. To prevent NEJ attachment to tips, use tips treated with silicone or PBS-5% FBS.
19. In our hands, NEJs cultured in RPMI-1640 without fetal bovine serum can be maintained for at least 1 week with >80% viability. For longer periods, NEJ should be cultured as described in Chapter 6.
20. Parasite should be collected in the same bile to prevent the presence of flukicidal compounds that can interfere with parasite viability.
21. Temperature is critical to maintain parasites with high viability and motility. This is crucial when you want to culture adult parasites in the lab.
22. Parasites collected from the bile duct show a green-dark gut. Viable parasites regurgitate gut content and become clear (yellowish brown). This wash step is crucial when you want to study parasite protein (e.g., excretion–secretion product or somatic proteins) and to avoid host protein contamination.

Acknowledgments

We would like to thank Lic. Federico Fossa and Dr. Nicolas Dell'Oca for sharing images related to egg culture and snail infection. We also would like to thank Dr. Gabriel Rinaldi and Dr. Gabriela Maggioli for sharing images related to metacercarial excystment and adult parasite.

References

1. Carmona C, Tort JF (2017) Fasciolosis in South America: epidemiology and control challenges. J Helminthol 91(2):99–109
2. Malatji MP, Pfukenyi DM, Mukaratirwa S (2019) Fasciola species and their vertebrate and snail intermediate hosts in east and southern Africa: a review. J Helminthol 94:1–11
3. Ashrafi K et al (2014) Fascioliasis: a worldwide parasitic disease of importance in travel medicine. Travel Med Infect Dis 12(6 Pt A):636–649
4. Jaja IF et al (2017) Financial loss estimation of bovine fasciolosis in slaughtered cattle in South Africa. Parasite Epidemiol Control 2 (4):27–34
5. Jefferies HS, Dawes B (1960) Elucidation of the life-cycle of *Fasciola hepatica*. Nature 185:331–332
6. Kendall SB, Ollerenshaw CB (1963) The effect of nutrition on the growth of *Fasciola hepatica* in its snail host. Proc Nutr Soc 22:41–46
7. Nari A et al (1986) Estudio preliminar sobre el desarrollo de Limnaea viatrix D´ Orbigny (1835) en condiciones controladas de temperatura y humedad. Veterinaria 95:13–17
8. Kendall S (1953) The life history of Limnaea truncatula under laboratory conditions. J Helminthol 27:17–28
9. Sukhdeo M, Sukhdeo S (2004) Trematode behaviours and the perceptual worlds of parasites. Can J Zool 82:292–315
10. Nari A et al (1983) Efecto de la temperatura en el desarrollo de Fasciola hepática en su huésped intermediario Limnaea viatrix D´ Orbigny (1835). Veterinaria 19:36–39

Chapter 2

Microscopical Techniques to Analyze the Hepatic and Peritoneal Changes Caused by *Fasciola hepatica* Infection

M. T. Ruiz-Campillo, V. Molina-Hernández, J. Pérez, I. L. Pacheco, R. Pérez, A. Escamilla, F. J. Martínez-Moreno, A. Martínez-Moreno, and R. Zafra

Abstract

The helminth parasite *Fasciola hepatica* modulates the host immune response at early stages of infection (Rodríguez et al., PLoS Negl Trop Dis 9:e0004234, 2015; Vukman et al., J Immunol 190:2873–2879, 2013). Nevertheless, little is known about the cell composition of the peritoneal fluid at these early stages of infection.

In this chapter, we describe a method to perform peritoneal lavages and to recover peritoneal fluid from sheep experimentally infected and noninfected with *F. hepatica* at early stages of infection. In addition, with the aim to characterize the peritoneal fluid immune cell phenotype, we describe a procedure to obtain the total leukocyte count, the differential leukocyte count and the preparation and storage of peritoneal fluid smears, together with the application of an immunocytochemical technique and an automatic method to count the immunoreactive cells. Finally, the present protocol describes the evaluation of the gross and the histopathological lesions together with the immunohistochemical analysis of the hepatic tissue.

Key words *Fasciola hepatica*, Macrophages, Peritoneal fluid, Sheep, Immune response, Immunohistochemistry, Leukocyte count, Peritoneal lavages

1 Introduction

F. hepatica is able to modulate the host immune response at early stages of infection. However, little is known about cells subpopulations present in peritoneal fluid and their function [1, 2]. The search for an effective method to obtain peritoneal fluid from animals experimentally infected with *F. hepatica* and its subsequent application and evaluation in the laboratory have been pursued by many authors [3]. Firstly, this method was described in rats experimentally infected with *F. hepatica* by carrying out cavity lavage using sterile phosphate buffered saline (PBS) containing 6 mM ethylenediaminetetraacetic acid (EDTA). This fluid was centrifuged at 252 × *g* for 10 min and the cell pellet resuspended in sterile

Martin Cancela and Gabriela Maggioli (eds.), *Fasciola hepatica: Methods and Protocols*, Methods in Molecular Biology, vol. 2137, https://doi.org/10.1007/978-1-0716-0475-5_2, © Springer Science+Business Media, LLC, part of Springer Nature 2020

Roswell Park Memorial Institute medium (RPMI) to evaluate antibody-dependent cell-mediated cytotoxicity by resident peritoneal lavage cell populations using enzyme-linked immunosorbent assay (ELISA) [3]. Then, a protocol for peritoneal fluid recovery was described in rats to estimate cell phenotypes within the peritoneal cavity during *F. hepatica* migration using flow cytometry. In this protocol, a warm solution (37 °C) of Dulbecco's phosphate buffered saline (DPBS), pH 7.2 and 2 units mL^{-1} of heparin is injected in the peritoneal cavity of rats. After gentle abdominal massaging for 1 min, peritoneal fluid is collected by abdominal incision. The suspended cells are washed and quantified using a hemocytometer, and cell viability is determined by trypan blue exclusion method [4]. Eventually, a method to collect peritoneal fluid from sheep experimentally infected with *F. hepatica* was developed by our group [5]. In these studies, the Diff-Quick-stained smear method for the differential peritoneal cell count was used. Nevertheless, the aim of this chapter is to describe a different, innovative, and precise method to perform a differential cell count based on two main steps: (1) immunocytochemistry with anti-human CD68, which provides a brownish color in the macrophage cytoplasm and (2) eosin dye for staining the cytoplasmic granules in eosinophils. Finally, protocols for gross and microscopical liver evaluation are included.

2 Materials

2.1 Recovery of Peritoneal Fluid

1. PBS: 7.2 g sodium chloride, 0.43 g potassium dihydrogen phosphate, and 1.48 g disodium hydrogen phosphate are added to 1 L of dH_2O in a flask. Store at 4 °C (*see* **Note 1**).
2. T-61 euthanasia solution: 800–1200 mg embutramide, 200–300 mg mebezonio ioduro, 20–30 mg tetracaine hydrochloride for 50 kg weight (4–6 mL/50 kg weight, approximately).
3. PBS-heparin solution: 9500 IU heparin in sterile PBS.
4. Electric razor.
5. 10% polyvinylpyrrolidone–iodine: commercial standard solution.
6. Blunt scissors.
7. Three 20 mL syringes and a 40 cm plastic cannula.
8. Two 50 mL conical tubes and a bucket with chipped ice.

9. Erythrolysis buffer: 8.26 g ammonium chloride, 1 g potassium bicarbonate, 0.037 g EDTA, and 1 L dH_2O. Store at room temperature (*see* **Note 2**).
10. Surgical scalpel handle and blade.

2.2 Total Peritoneal Cell Count

1. 0.4% trypan blue dye solution: 40 mg of trypan blue powder in 100 mL of PBS.
2. 1.5 mL tube.
3. Optical microscope.
4. Neubauer counting chamber (hemocytometer).
5. Cover glass.
6. 10–100 μL pipette with disposable tips.
7. Hand tally counter.

2.3 Differential Cell Count Using Immunocytochemistry and Hematoxylin and Eosin Counterstain

1. Vectabond-treated slides (Vector Laboratories, CA, USA).
2. Slide staining rack.
3. Acetone.
4. −80 °C freezer.
5. Diamond pencil.
6. Blocking endogenous peroxidase solution: 3% hydrogen peroxide in PBS. In a crystal staining jar, add 196 mL PBS and 6 mL 33% hydrogen peroxide.
7. Blocking nonspecific reactions solution: 10% of normal goat serum (MP Biomedicals, Ohio, USA) diluted in PBS.
8. Primary antibody solution: antibody anti-human CD68 (M0718, Dako, Glostrup, Denmark) diluted 1:400 in 10% normal goat serum.
9. PBS-Tween 20 (PBST) solution: 0.5% Tween 20 in PBS. Mix 10 mL Tween 20 with 2 L PBS in a flask with a magnetic stirrer (*see* **Note 3**).
10. Secondary antibody solution: biotinylated goat anti-mouse immunoglobulin serum (Dako) diluted 1:50 in 10% normal goat serum.
11. 1:50 avidin–biotin–peroxidase complex.
12. Chromogen solution: Vector novaRED Peroxidase Substrate Kit.
13. 10% Harris hematoxylin in distilled water.
14. Eosin.
15. 70% alcohol, 96% alcohol, 100% alcohol, and xylol.
16. Coverslips.

17. Quick-hardening mounting medium Eukitt (Freiburg, Germany).
18. 10% formalin in PBS.

2.4 Pathological Examination

1. Paraffin wax.
2. Microtome.
3. Hematoxylin and eosin stain.
4. Acid alcohol: 1% hydrochloric acid in 70% reagent grade alcohol.
5. Image Pro-Plus 6.0 Software (Media Cybernetics, Silver Spring, MD, USA).

3 Methods

3.1 Recovery of Peritoneal Fluid

1. Euthanasia is applied to the animals by an intravenous injection of T-61 euthanasia solution in the jugular vein. Immediately thereafter, the ventral aspect of the abdomen is shaved and disinfected with 10% polyvinylpyrrolidone–iodine (*see* **Note 4**).
2. A 2 cm incision cranially to the umbilicus scar is made on the skin over the white line and subcutaneous tissue is dissected with blunt scissors to avoid bleeding.
3. A 40 cm length cannula is inserted in the abdominal cavity and connected with a syringe to inject 60 mL of PBS-heparin solution, previously heated at 37 °C (*see* **Note 5**).
4. The abdominal cavity is softly massaged for 1 min.
5. Between 40 and 50 mL of peritoneal fluid can be recovered using the syringe connected to the cannula. Immediately, the cell suspension collected is transferred into a 50 mL conical tube and placed in chipped ice.

3.2 Total Peritoneal Cell Count

1. The total number of viable peritoneal cells is determined by a 1:10 dilution of trypan blue cell suspension prepared in a 1.5 mL tube. The total peritoneal cell count will be carried out using a Neubauer counting chamber (*see* **Note 6**).
2. The Neubauer is a counting-chamber device that consists of a thick crystal slide (0.1 mm depth, 0.0025 m^2). This means that the space between the chamber and the coverslip is 0.1 mm and the smallest square on the grid has an area of 0.0025 mm^2. The Neubauer counting chamber has two counting areas that can be loaded independently.
3. Put a clean glass cover on the Neubauer counting-chamber central area. Use a flat surface to place the chamber, like a table or a workbench.

4. Put a disposable tip at the end of the pipette.
5. Take the trypan blue cell solution prepared in **step 1** and mix well (pipet up and down).
6. Adjust the pipette to take 10 μL. Push the pipette plunger slowly up to the first resistance point of the pipette. Introduce the pipette tip in the cell dilution and gently release the plunger to load the amount of liquid adjusted.
7. Remove the pipette from the dilution and bring it to the Neubauer counting chamber. When the pipette is loaded, it must always be held in vertical position.
8. Place the pipette tip close to the glass cover edge, right at the center of the Neubauer counting chamber.
9. Release the plunger slowly watching how the liquid enters the chamber uniformly, being absorbed by capillarity. In the case of bubble formation or glass cover displacement, repeat the operation.
10. With the microscope, using a 4× objective, identify the nine main squares of the counting chamber delimited by three lines each.
11. Change to 10× objective and focus one of the nine main squares. The counting is performed in the area delimited by three lines. Cells that touch the upper and left border are counted, while cells that touch the right and lower border are not counted.
12. Perform the counting procedure in the four squares placed at the corners. Use a hand tally counter for that purpose.
13. Count the number of cells for each main corner square. Then, obtain the average of the four main squares. This number should be multiplied by 10^5 to obtain the number of cells per mL (*see* **Note 7**).

3.3 Differential Cell Count Using Immunocytochemistry and Hematoxylin and Eosin Counterstain

1. Centrifuge peritoneal fluid at 252 × *g* for 10 min and discard the supernatant. Resuspend the cell pellet in 5 mL erythrolysis buffer and incubate for 10 min at room temperature. Centrifuge at 252 × *g* for 5 min and discard supernatant.
2. Add 1 mL fresh and sterile PBS to the cell pellet. Pipet up and down and mix thoroughly.
3. Pipet 10 μL of this dilution. Place the drop on a very clean slide. Hold a second slide making sure that its edge is in contact with the bottom slide (Fig. 1).
4. Drag the top slide edge back to soak it with the drop. This will spread by capillarity.

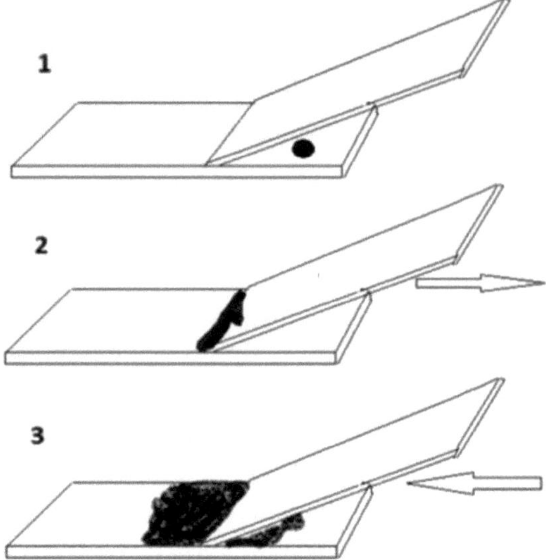

Fig. 1 Place a small drop of 10 μL on the slide (1). Maintain firm contact with the bottom slide and drag the top slide edge back to stretch the drop with it (2). Maintain firm contact with the bottom slide and push the top slide in one swift movement to produce the smear (3)

5. Maintain firm contact with the bottom slide and push the top slide in one swift movement to make the smear. Extensions are performed in Vectabond-treated slides (*see* **Note 8**).
6. Place the smears in a slide staining rack and immerse this in a crystal staining jar containing 200 mL of pure cold acetone for 5 min. Let them air-dry and store at −80 °C.
7. Take the peritoneal fluid smears out of the −80 °C freezer and let them air-dry. Use a diamond pencil to carve a circle around the sample and an identification symbol in one of the corners of the slide to help us identify the samples during the immunocytochemical process.
8. Place the slides in a slide staining rack. Immerse the slide staining rack in a crystal staining jar containing 200 mL of pure cold acetone for 10 min. Take out the slide staining rack and let the sections air-dry at room temperature.
9. Moisten two pieces of laboratory paper and place them in each of the two gaps of a 100-slide box. The smears will be placed in horizontal position with the sample faced up inside the slide box (*see* **Note 9**).
10. Block endogenous peroxidase activity by incubation with blocking endogenous peroxidase solution for 15 min in darkness (*see* **Note 10**).
11. Wash sections once in PBS for 10 min.

12. Incubate in blocking nonspecific reactions solution for 30 min at room temperature (*see* **Note 11**).
13. To prepare the primary antibody solution, dilute antibody anti-human CD68 solution 1:400 in 10% normal goat serum (*see* **Note 12**).
14. Carefully dry the surroundings of the carved-marked sample with laboratory paper and place them back in the slide box.
15. Apply primary antibody solution; 100 μL should be enough to cover the sample and keep it wet.
16. Carefully close the slide box containing these smears and leave it overnight at 4 °C. The remaining PBS is also kept in the fridge (*see* **Note 13**).
17. Let the samples in the slide box and the PBS gain the room temperature (1 h).
18. Rinse three times, 5 min each in PBST. Gently remove the excess of liquid from the periphery of the sample outside the carved line with a clean paper and place it in the slide box.
19. Prepare secondary antibody solution: a biotinylated goat anti-mouse immunoglobulin serum diluted 1:50 in 10% normal goat serum is applied to the sections. Incubate the samples for 45 min at 37 °C (*see* **Note 14**).
20. Rinse three times, 5 min each in PBST. Gently remove the excess of liquid from the periphery of the sample outside the carved line with a clean paper and place it in the slide box.
21. Prepare the avidin–biotin–peroxidase complex (1:50) and apply to the sections for 1 h at room temperature in darkness (*see* **Note 15**).
22. Rinse three times, 5 min each in PBST. Gently remove the excess of liquid from the periphery of the sample outside the carved line with a clean paper and place it in the slide box.
23. Incubate samples with chromogen solution diluted following the manufacturer's instructions.
24. Rinse in water and counterstain the slides with 10% Harris hematoxylin in distilled water for 5 min. Rinse in water.
25. Stain the slides in eosin for 1 min. Rinse in water.
26. Dehydrate doing quick passes through 70% alcohol, 96% alcohol, 100% alcohol, and xylol. Mount the slides with a coverslip and a drop of Eukitt.
27. Under the optical microscope, a total of 200 cells per animal is counted and the percentage of eosinophils, lymphocytes, and macrophages obtained.

3.4 Pathological Examination

1. All sheep are subjected to necropsy.
2. Image Pro-Plus 6.0 Software is used to conduct morphometrical studies.

3.4.1 Gross Pathology

1. The diaphragmatic and visceral surfaces of the liver are photographed.
2. Gross pictures of livers are used to measure the area of the damaged hepatic parenchyma (fibrosis, chronic tracts).
3. The perimeter of the diaphragmatic and visceral aspects of the liver is delineated and the total area obtained.
4. The damaged areas (scars, fibrosis, hemorrhages) are delineated and the total damaged area in the visceral and diaphragmatic aspects is obtained.
5. The mean value (SD) for each group is calculated (*see* **Note 16**).

3.4.2 Histopathology

1. Tissue samples are collected from the left and right hepatic lobes with a scalpel and surgical tweezers. For each animal, four hepatic tissue samples from affected areas are taken.
2. Tissues are fixed in 10% formalin and embedded in paraffin wax. Small samples are cut with a microtome (4 mm) and placed on a clean slide.
3. Tissue sections are stained with hematoxylin and eosin (HE) (*see* **Note 17**).
4. Four tissue sections per animal are evaluated independently by two pathologists to assess the severity of histopathological hepatic lesions (*see* **Note 18**).
5. A morphometric study is conducted on HE-stained tissue sections to obtain two parameters (*see* **Note 19**). Eight randomly selected low-power (×25) photomicrographs per animal are used, each photomicrograph measuring an area of 8.5 mm^2 (i.e., 68 mm^2 are examined per animal).
6. For the first parameter, the perimeter of the portal areas and all damaged areas are delineated and the area is obtained. For the second parameter, the outer perimeter of bile ducts is delineated. Results are expressed in percentage values (mean SD) per group (*see* **Note 20**).

4 Notes

1. Use a previous amount of water in the flask (approximately 0.5 L) to dissolve the reagents easily and more quickly. Adjust pH of the solution to 7.2.

2. Use a previous amount of water in the flask (approximately 0.5 L) to dissolve the reagents easily and more quickly.
3. Add 5 mL of Tween 20 per liter of PBS with a 10 mL serological pipette and a pipette aspirator. The remaining PBS is 1.6 L the second day of the immunocytochemistry technique, so 8 mL of Tween 20 is added. Because Tween 20 has a gooey consistency, this should be added to PBS while the magnetic stir is on in order to mix them properly.
4. The polyvinylpyrrolidone iodine is applied with a cotton compress once the ventral region of the abdomen has been shaved with the electric razor. The experiments are approved by the Bioethics Committee of the University of Cordoba (No.1118) and conducted in accordance with European (2010/63/UE) and Spanish (RD 1201/2005) directives on animal experimentation.
5. The PBS-heparin solution must be prepared and kept in the stove the day before. It is recommended to prepare the three 20 mL syringes containing the PBS-heparin in advance. The last syringe used to inject the PBS-heparin can be used to recover peritoneal fluid after the massage.
6. To prepare the suspension of cells with trypan blue dye, use 90 μL of trypan blue and 10 μL of peritoneal cell suspension.
7. The concentration will be the total number of cells counted $\times$ 10^5/number of squares. The number of cells will be the sum of all the counted cells in all squares counted. The volume will be the total volume of all the squares counted.

   Since the area of one corner square is $0.1\ \text{cm} \times 0.1\ \text{cm} = 0.01\ \text{cm}^2$ of area counted and since the depth of the chamber is $0.1\ \text{mm} = 0.01\ \text{cm}$

   $0.01\ \text{cm}^2 \times 0.01 = 0.0001\ \text{cm}^3 = 0.0001\ \text{mL}$

   Because a dilution of 1:10 has been done, 0.0001 mL should be multiplied by 0.1:

   Final concentration: number of cells $\times$ 10^5/number of squares.
8. Vectabond is a reagent for tissue section adhesion designed to significantly increase the adherence of both frozen and paraffin-embedded tissue sections and cell preparations to glass slides and coverslips.
9. The preparation of a moisture chamber avoids samples to dry and the subsequent occurrence of false positives.
10. Endogenous peroxidase will react with the substrate solution, leading to false positives. This nonspecific background can be significantly reduced by pretreatment of the sample with hydrogen peroxide before incubation with the secondary

antibody. Hydrogen peroxide is diluted in methanol or in PBS. Because some cell surface protein markers are very sensitive to methanol, especially in frozen sections, using hydrogen peroxide in PBS is suggested.

11. It is effective in reducing nonspecific binding of proteins and prevents the secondary antibody from cross-reacting with endogenous immunoglobulins in the tissue. The use of normal serum block before the application of primary antibody also eliminates Fc receptor binding of both the primary and secondary antibodies. Usually, the amount of dilution used for each slide is 100 μL. Thus, for each slide 10 μL of normal goat serum and 90 μL of PBS are required (10% normal goat serum). Because this quantity is needed per slide, it has to be adjusted considering the number of slides used per immunocytochemistry.

12. To avoid air drying of the sections, it is recommended to prepare the primary antibody in advance. 10% normal goat serum is added in a clean 1.5 mL microtube and then the anti-human CD68 antibody (following the 1:400 dilution).

13. It is important to be careful when placing the slide box inside the fridge to avoid the sections from moving or tilting.

14. The secondary antibody dilution should be prepared in advance in order to avoid the sections to air-dry.

15. The avidin–biotin complex staining method serves as a technique to amplify the target antigen signal due to the multiple binding opportunities between the avidin and biotinylated antibodies (bound to the antigen). The ABC complex consists of two reagents (A and B) which must be mixed using the following formula:

    2 μL reagent A + 2 μL reagent B + 96 μL PBS = 100 μL

    100 μL is needed to cover one sample in the section, so the numbers in the formula must be multiplied by the number of sections used in the current immunocytochemistry.

16. Statistical analysis is carried out with GraphPad 7.0 software (GraphPad Software Inc., San Diego, CA, USA). The Kolmogorov–Smirnov test is applied to decide whether distributions were parametric. A comparison between pairs of groups is made with the Mann–Whitney U test. $P < 0.05$ is considered significant. Correlation between fluke burdens and gross hepatic lesions is assessed by the Spearman correlation test for nonparametric distributions, with $P < 0.05$ again being considered significant.

17. Air-dry the tissue samples. Stain nuclei with the hematoxylin (30 min). Rinse in running tap water (3 min). Differentiate with 1% acid alcohol for 4 s (198 mL 96° ethanol and 2 mL

HCl). Rinse in running tap water (5 min). Stain with eosin for 10 min. Rinse in running tap water until totally clean. Dehydrate, clear, and mount.

18. The pathological parameters fibrous perihepatitis, bile duct hyperplasia, portal inflammatory infiltrate of lymphocytes and plasma cells, portal fibrosis granulomas and infiltrate of eosinophils) are evaluated as follows: 0, absent; 1, mild; 2, moderate; 3, severe; and 4, very severe.
19. The parameters obtained are (1) the area of portal spaces and area of damaged tissue (chronic tracts, cholangitis with inflammatory infiltrate composed by eosinophils, lymphocytes and plasma cells, fibrosis and granulomas); and (2) area of the bile ducts.
20. Cell counting and statistical analysis: Immunoreactive cells are counted in 10 fields of 0.2 mm^2, randomly selected in damaged hepatic areas, with the Image-Pro Plus 6.0 software. Statistical analysis is carried out with GraphPad 7.0 software (GraphPad Software Inc., San Diego, CA, USA). The Kolmogorov–Smirnov test is applied to decide whether distributions were parametric. A comparison between pairs of groups is made with the Mann–Whitney U test; $P < 0.05$ is considered significant. The correlation between fluke burdens and gross hepatic lesions is assessed by the Spearman correlation test for nonparametric distributions, with $P < 0.05$ again being considered significant.

References

1. Rodríguez E, Noya V, Cervi L, Chiribao ML, Brossard N, Chiale C, Carmona C, Giacomini C, Freire T (2015) Glycans from *Fasciola hepatica* modulate the host immune response and TLR-induced maturation of dendritic cells. PLoS Negl Trop Dis 9(12):e0004234
2. Vukman KV, Adams PN, Metz M, Maurer M, O'Neill SM (2013) *Fasciola hepatica* tegumental coat impairs mast cells' ability to drive Th1 immune responses. J Immunol 190 (6):2873–2879
3. Jedlina L, Kozak-Ljunggren M, Wedrychowicz H (2011) In vivo studies of early, peritoneal, cellular and free radical response in rats infected with *Fasciola hepatica* by flow cytometric analysis. Exp Parasitol 128:291–297
4. Piedrafita D, Parsons JC, Sandeman RM, Wood PR, Estuningsih SE, Partoutomo S, Spithill TW (2001) Antibody-dependent cell mediated cytotoxicity no newly excysted juvenile *Fasciola hepatica* in vitro is mediated by reactive nitrogen intermediates. Parasite Immunol 23:473–482
5. Zafra R, Pérez-Écija RA, Buffoni L, Pacheco IL, Martínez-Moreno A, LaCourse EJ, Perally S, Brophy PM, Pérez J (2013a) Early hepatic and peritoneal changes and immune response in goats vaccinated with a recombinant glutathione transferase sigma class and challenged with *Fasciola hepatica*. Res Vet Sci 94:602–609

Chapter 3

Isolation of Secreted and Tegumental Surface Proteins from *Fasciola hepatica*

Eduardo de la Torre-Escudero and Mark W. Robinson

Abstract

Proteins secreted by, or displayed on the surface tegument of, trematodes have key functions in the host–parasite interaction. As such, they are often leading targets for diagnostic tests or vaccine candidates. Here we describe methods for the isolation and analysis of soluble secreted proteins (i.e., the secretome) released during in vitro culture of adult *Fasciola hepatica*. We also describe two methods for the enrichment of proteins displayed on the outer tegumental surface of *F. hepatica*. These approaches enable downstream identification of the isolated proteins by mass spectrometry-based proteomics.

Key words *Fasciola hepatica*, Secretome, Excretory–secretory products, Tegument, Host–parasite interaction, Proteomics

1 Introduction

The continued technological advance of mass spectrometry-based proteomics, together with the recent availability of comprehensive genome and transcriptome datasets for *Fasciola hepatica* [1], has significantly enhanced our ability to identify proteins expressed by this species. Of particular interest to study are those proteins that operate at the interface between the parasite and the host to ensure the establishment of infection and the long-term survival of the parasite, for example, via modulation of the host immune response. This interface comprises those proteins excreted and secreted by the worms (either soluble secreted proteins or those packaged within extracellular vesicles) and those expressed on the outer surface of the tegument [2–5]. As a result of these proteomics studies, several promising vaccine candidates have been identified and tested in the past decade, but their protection efficacy has not yet reached that required for commercialization [6]. Thus, further work to better understand how *F. hepatica* interacts with its mammalian host at an immunological level is required as well as further identification of appropriate vaccine targets. In this chapter, we

Martin Cancela and Gabriela Maggioli (eds.), *Fasciola hepatica: Methods and Protocols*, Methods in Molecular Biology, vol. 2137, https://doi.org/10.1007/978-1-0716-0475-5_3, © Springer Science+Business Media, LLC, part of Springer Nature 2020

describe a series of methods for the isolation, enrichment, and downstream identification of *F. hepatica* proteins that may be involved in the host–parasite interaction. We start by describing how to culture flukes in vitro to collect their secretions followed by a series of methods to concentrate/precipitate secreted proteins for subsequent proteomics analysis [2, 7]. Finally, we describe two complementary approaches aimed at identifying those proteins displayed specifically on the outer surface of the tegument [8–10].

2 Materials

Unless stated otherwise, all steps involving live worms must be performed at 37 °C; therefore, solutions must be prewarmed to ensure optimal parasite survival in vitro. Use of sterile buffers and flasks is recommended (*see* **Note 1**).

2.1 Collection of F. hepatica Excretory–Secretory Proteins In Vitro

1. Prepare enough sterile PBS supplemented with 0.1% glucose to collect the flukes from the abattoir. Prewarm PBS and carry it in thermos flasks to maintain its temperature at 37 °C (*see* **Note 2**).
2. Use RPMI 1640 medium containing 2 mM L-glutamine, 30 mM HEPES to culture the flukes (*see* **Note 3**). Before use, supplement with 0.1% (w/v) glucose and add antibiotics at a final concentration of 100 U/mL penicillin and 100 μg/mL streptomycin.

2.2 Filter Device Concentration of Excretory–Secretory Products

1. Amicon® Ultra-15 centrifugal filter device (Millipore) with a 3-kDa molecular weight cutoff (*see* **Note 4**).

2.3 Acetone Precipitation from Large Volumes

1. Prechill a stock of HPLC grade acetone at −20 °C for this method.
2. Suitable acetone-resistant polypropylene tubes should be used.

2.4 Methanol–Chloroform Precipitation from Small Volumes

General purpose reagents (GPR) grade chemicals can be used for this method.

2.5 In-Gel Trypsin Digestion

1. Equilibration solution. Prepare 100 mM ammonium bicarbonate solution by dissolving 345 mg of NH_4HCO_3 in 50 mL of dH_2O.
2. Wash solution. Prepare a 50% acetonitrile–50 mM NH_4HCO_3 solution by adding equal volumes of acetonitrile and 100 mM NH_4HCO_3 stock solutions.

3. Reduction and alkylation solution. Make up 1 mL of 100 mM NH_4HCO_3 containing 5 mM tributylphosphine (TBP) and 20 mM acrylamide solution, by adding 950 μL of 100 mM NH_4HCO_3, 20 μL of 1 M acrylamide and 30 μL of TBP solution.
4. Trypsin solution. Add 100 μL of 1 mM HCl to a 20 μg vial of trypsin (Proteomics grade trypsin) and vortex (*see* **Note 5**). Subsequently, add 900 μL of 50 mM NH_4HCO_3 to this to make a final concentration of 20 ng/μL trypsin at pH 8.

2.6 Biotinylation of F. hepatica Tegumental Surface Proteins

1. Membrane impermeable biotin (EZ-Link Sulfo-NHS-LC-Biotin).
2. Biotin solution: let the biotin warm to room temperature before weighing it out (*see* **Note 6**). Prepare 0.5 mg/mL of biotin solution (~1 mM) in PBS, pH 7.4 (0.005 g up to 10 mL PBS).
3. Quench solution, 50 mM Tris–HCl in PBS, pH 7.4.
4. EDTA-free proteinase inhibitor cocktail, use one tablet per each 10 mL of solution.

2.7 Protein Extraction and Streptavidin Pull-Down of the Biotinylated Proteins

1. Tissue homogenizer (such as SHM1, Stuart) and appropriate probe for processing volumes up to 5 mL.
2. High capacity streptavidin–agarose resin.
3. Solubilization buffer: prepare 2% (w/v) SDS, 50 mM Tris, 10 mM EDTA in PBS (pH 6.8) and add one tablet of complete EDTA-free proteinase inhibitor cocktail per each 10 mL of solution.
4. Wash buffer: prepare a dilution of 0.1% SDS (w/v) in PBS (pH 7.4).
5. Digestion buffer: prepare 50 mM NH_4HCO_3, pH 8.0 (197.5 mg of NH_4HCO_3 up to 50 mL of dH_2O).

2.8 Soft Trypsin Digestion of F. hepatica

1. Fresh and methanol treated parasites.
2. Dithiothreitol (DTT) (Thermo Scientific).
3. Iodoacetamide (IAM), single use (Thermo Scientific). Prepare fresh right before use and protect from light.

3 Methods

Although most of the procedures included in this chapter focus on adult worms, they may be applied to other life stages of the parasite with minor modifications.

3.1 Collection of *F. hepatica* Excretory–Secretory Proteins In Vitro

1. Wash the adult worms extensively in prewarmed PBS to remove any contaminant blood and bile (*see* **Note 7**).
2. Preincubate the worms for 30 min in prewarmed RPMI 1640 medium to allow them to void their gut contents before collecting the secretions.
3. Count and separate the flukes into sterile flasks and culture them at a concentration of 1 mL/fluke in RPMI medium at 37 °C for 5 h (*see* **Notes 8** and **9**).
4. Collect the culture medium that contains *F. hepatica* excretory–secretory (ES) proteins and spin at 300 × *g* at 4 °C for 10 min to remove eggs and debris.
5. Carefully take the supernatant to a new tube and spin at 2000 × *g* at 4 °C for 30 min to remove dead cells.
6. Pool the culture medium and proceed with one of the following techniques according to your experimental requirements (*see* **Note 10**).

3.2 Filter Device Concentration of Excretory–Secretory Products

1. Prerinse an Amicon® Ultra-15 centrifugal filter device with a 3-kDa molecular weight cutoff, with ddH_2O or buffer right before use (*see* **Note 11**).
2. Add 12 mL of the culture medium (ES) to the centrifugal filter device.
3. Place capped filter device into the centrifuge rotor and balance it with a similar device.
4. Spin the culture medium at 5000 × *g* maximum for approximately 15 min and check the volume of the flow-through to determine the optimal spin time for your sample (generally between 15 and 60 min).
5. Top up the centrifugal filter device with ES medium and repeat the process until all the sample has been concentrated to a final concentration of 1 mg/mL (routinely, the volume can be reduced to 200–400 μL).
6. Recover the concentrated ES from the filter device with a pipette immediately after centrifugation to ensure optimal recovery.
7. Aliquot and store the concentrated culture medium at −80 °C for further analysis.

3.3 Acetone Precipitation from Large Volumes of ES

1. Add 5 volumes of −20 °C cold acetone to the protein sample (e.g., 1 mL sample + 5 mL acetone). Make sure that the tube is acetone safe and vortex to mix thoroughly.
2. Incubate for 1 h at −20 °C.
3. Centrifuge the sample at 13,000–15,000 × *g* for 10 min at 4 °C.

4. Decant the supernatant and air-dry the pellet for 5–10 min. Do not overdry the pellet, as this may make solubilization difficult.
5. Solubilize the pellet in an appropriate buffer for your downstream applications.

3.4 Methanol–Chloroform Precipitation from Small Volumes of ES

1. Add 500 μL of methanol–chloroform (4:1) to 400 μL of ES (in RPMI).
2. Add 300 μL of dH_2O and vortex for 10 s.
3. Spin at 12,470 × *g* for 2 min at room temperature.
4. Discard the upper methanol–water phase (protein is at the interphase).
5. Add 500 μL of methanol and spin at 13,000 rpm for 5 min at room temperature.
6. Remove the supernatant and briefly air-dry the pellet.
7. Solubilize the pellet in an appropriate buffer for your downstream application.

3.5 In-Gel Trypsin Digestion

Take steps to avoid keratin contamination—wipe down surfaces with methanol; work in a hood if possible; and wear gloves and lab coat.

1. Excise stained bands or spots from a gel using a clean scalpel on a glass plate and place in a clean 0.5 mL capped tube. Large gel pieces should be cut into smaller squares (1 mm^3). Gel pieces can be stored at −20 °C.
2. Wash the gel pieces briefly with 100 μL of equilibration solution (100 mM NH_4HCO_3), with vortexing, to make sure that the gel pieces are at the correct pH.
3. To de-stain Coomassie blue-stained gel pieces, add 200 μL of wash solution (50% acetonitrile–50 mM NH_4HCO_3) and vortex well. Remove the liquid and repeat until the color is gone (*see* **Note 12**).
4. Remove the excess liquid and dehydrate the gel pieces with 200 μL of 100% acetonitrile for 5 min. Remove the acetonitrile and allow to air-dry (by leaving the caps open) for 10 min. The gel pieces should be noticeably shrunken and are often white in color.
5. Reduce and alkylate the samples by incubating the gel pieces for 90 min at room temperature with 80 μL of 100 mM NH_4HCO_3 containing 5 mM tributylphosphine (TBP) and 20 mM acrylamide.
6. Wash in 200 uL of 100 mM NH_4HCO_3 for 5 min with vortexing.
7. Give 2 × 5 min washes in 200 μL of wash solution (50% acetonitrile–50 mM NH_4HCO_3) with vortexing.

8. Repeat **step 4**.
9. Add 30 μL of 2 ng/μL sequencing grade trypsin and incubate at 4 °C for 30 min.
10. Add approximately 30 μL of 50 mM NH_4HCO_3 to cover the gel pieces and incubate overnight at 37 °C.
11. To extract the peptides, remove the excess liquid and retain in a clean tube. Add 35 μL of 50% acetonitrile–2% formic acid to the gel pieces and incubate at 30 °C for 20 min with frequent vortexing.
12. Transfer the excess liquid to a new tube and repeat **step 11**.
13. Combine all extracts and concentrate using a vacuum concentrator (at 30 °C) until the sample volume reaches approx. 15 μL. Peptide samples can be stored at −20 °C prior to analysis by mass spectrometry.

3.6 Biotinylation of *F. hepatica* Tegumental Surface Proteins

1. Wash and preincubate the flukes as indicated in Subheading 3.1 (**steps 1** and **2**) to ensure that they empty their gut contents (*see* **Note 13**).
2. Incubate the parasites with biotin labeling solution for 30 min at 4 °C with gentle agitation in an orbital shaker (*see* **Note 14**).
3. Remove the labeling solution and neutralize the unbound biotin with quenching solution (50 mM Tris–HCl in PBS) for 15 min at 4 °C.
4. Wash the worms extensively with PBS, remove the liquid and continue with the protein extraction as indicated below (Subheading 3.7). At this point, samples may be stored before further processing, add a small volume of PBS containing protease inhibitors (enough to cover the worms) and snap-freeze them in liquid nitrogen.

3.7 Protein Extraction and Streptavidin Pull-Down of the Biotinylated Proteins

Homogenization of the worms and protein extraction with high concentration of detergents (*see* **Note 15**)

1. Resuspend the worms in 40 μL of solubilization buffer per mg of sample.
2. Homogenize the worms six times, 2 min each, at full power, keeping samples for 1 min on ice between homogenization steps. Avoid producing too much foam (*see* **Note 16**).
3. Sonicate the homogenate six times, 30 s each, at 40% intensity, and cool on ice for 30 s after each sonication step.
4. Incubate the homogenate at 95 °C for 20 min. Cool down to room temperature.
5. Centrifuge at 20,000 × *g* for 20 min at room temperature.

6. Collect the supernatant containing the solubilized proteins and dialyze overnight to reduce the concentration of SDS that might interfere with subsequent purification steps (*see* **Note 17**).

Affinity chromatography with streptavidin–agarose beads

1. Gently, suspend the streptavidin–agarose resin and pipet a volume enough to bind the proteins in your sample (typically 200 μL of settled beads for each 4 mg of total protein sample).
2. Centrifuge the resin at 500 × *g* for 1 min to remove the preservative solution, equilibrate the settled beads with PBS (2 × 1 mL) and then equilibrate with wash buffer (2 × 1 mL).
3. Add the protein extracts to the equilibrated beads and incubate for 1 h at room temperature with gentle agitation using an end-to-end rocker.
4. Collect the beads by centrifugation at 500 × *g* for 1 min and remove the supernatant containing unbound proteins (*see* **Note 18**).
5. Wash the beads three times with 600 μL of wash buffer and at least 4–6 times with 600 μL of PBS to remove detergent.
6. Wash the beads five times with 400 μL of digestion buffer (*see* **Note 19**).
7. Centrifuge the beads at 500 × *g* for 1 min and remove the supernatant. Add 2 mM dithiothreitol (DTT) in 50 mM NH_4HCO_3 and incubate at 60 °C for 20 min to reduce cysteine residues.
8. Centrifuge the beads at 500 × *g* for 1 min and remove the supernatant. Add 5 mM iodoacetamide (IAM) in 50 mM NH_4HCO_3 and incubate at room temperature for 30 min in the dark to alkylate sulfhydryl groups.
9. Neutralize excess IAM by adding 10 mM DTT and incubating for 30 min at room temperature.
10. Prepare the trypsin stock as indicated in Subheading 2.5, but adjusting the concentration to 40 ng/μL trypsin.
11. Add 300 μL of the digestion buffer plus 40 μL of the trypsin stock to the beads and incubate overnight at 37 °C with agitation (180 rpm in a shaking incubator).
12. Stop the reaction by adding trifluoroacetic acid (TFA) at a final concentration of 0.1%.
13. Transfer the beads to a filter device (Millipore, UFC30DV25) and centrifuge at 5000 × *g* for 1 min to collect the peptides (*see* **Note 20**).

3.8 "Soft" Trypsin Digestion of *F. hepatica* Newly Excysted Juveniles (NEJ)

1. Wash the parasites 3 × 1 min at 300 × *g* with 50 mM NH_4HCO_3.
2. Incubate the NEJs with 100 ng of sequencing grade trypsin, prepared as described in Subheading 2.5, in 50 μL of NH_4HCO_3 for 30 min at 37 °C.
3. Centrifuge at 300 × *g* and recover the supernatant that contains the peptides. Add DTT to a final concentration of 10 mM and incubate for 30 min at room temperature, to reduce the peptides.
4. Add IAA to a final concentration of 55 mM and incubate for 30 min at room temperature in the dark, to alkylate the peptides.
5. Stop the reaction with formic acid at 0.1% final concentration (for fresh parasites) or by adding 5 μL of 10% TFA (for methanol-fixed parasites).
6. Filter the supernatants containing the reduced and alkylated peptides through a 0.22 μm filter and evaporate to dryness using a vacuum concentrator.

4 Notes

1. Although it is not possible to collect worms from the abattoir under aseptic conditions, it is recommended to use sterile material during in vitro culture to minimize the risk of contamination with microorganisms.
2. Glucose hydrolyzes at high temperatures; therefore, it must be added to the PBS after sterilization.
3. The use of HEPES buffered medium is required to avoid extreme changes in pH due to the lactic metabolism of the worms.
4. Other MWCOs may be used but the MWCO of 3 kDa will ensure that most proteins are retained.
5. Trypsin may be stored for long periods at −20 °C if aliquoted in HCl. However, the pH needs to be adjusted after defrosting each aliquot by adding ammonium bicarbonate as described above.
6. EZ-Link sulfo-NHS-LC-biotin is easily hydrolyzed. Fresh solutions should always be prepared (typically no more than 10 min before use). As the reagent is moisture sensitive always equilibrate the vial to room temperature before opening.
7. *F. hepatica* eggs can be collected at this stage.

8. Use only the worms that show a clear aspect, that is, the ones with no pigment in their guts. Microscopic examination of flukes is encouraged so that dead/moribund flukes can be removed. We routinely keep culture time short (around 5 h) to keep collections physiologically relevant.
9. Avoid culturing flukes in large numbers, due to the lactate metabolism the pH of the medium will turn acidic and may harm the worms.
10. ES products may present very low protein concentration and need to be concentrated and the volume reduced before further analysis or storage. Depending on downstream analysis, ES material may be concentrated by filter device centrifugation to ensure that the biological activity of the proteins is retained. Alternatively, protein precipitation may be suitable for downstream applications (e.g., SDS-PAGE) that do not require preservation of biological activity.
11. Do not allow the membrane in Amicon® Ultra filter devices to dry out once wet. If you are not using the device immediately after rinsing, leave fluid on the membrane until the device is used.
12. Heavily stained bands may require numerous washes. However, if the color persists after 4–5 washes then proceed to the next step.
13. Handling of the worms should be kept to a minimum to avoid physical rupture of the tegument (e.g., transfer flukes between flasks by pouring them in culture media or use a soft paintbrush to manipulate them).
14. Include an equivalent control group of nonbiotinylated worms and process them alongside the biotinylated samples through the whole procedure.
15. The strong affinity of streptavidin for biotin ensures a highly efficient recovery of biotin-labeled proteins by using this extraction method. However, endogenously biotinylated proteins, present in other tissues, will be extracted alongside and may interact with the resin. Thus, it is important to include nonbiotinylated controls for this procedure. Alternatively, to avoid this, methods to detach the tegument in other trematode parasites have been described [11].
16. If necessary, centrifuge at 3000 × *g* for 1 min at 4 °C between the intervals to reduce frothing.
17. Take a 20 μL aliquot of each sample for protein determination with BCA protein assay. Protein extracts can be stored at −80 °C.

18. Take a 20 μL aliquot of the supernatant and the beads for SDS-PAGE and Western blot to assess the binding efficiency of the biotinylated proteins.
19. Before proceeding with the trypsin digestion, it is advisable to assess whether the affinity chromatography worked. Do not freeze the beads; keep them at 4 °C for trypsin digestion.
20. Peptides can be stored at −20 °C until analysis by mass spectrometry.

Acknowledgments

We thank Dr. Matt Padula (University of Technology, Sydney) and Drs. Ana Oleaga and Mar Siles Lucas (Institute of Natural Resources and Agrobiology, Salamanca) for original development of the in-gel trypsin digest and surface biotinylation protocols, respectively.

References

1. Cwiklinski K, Dalton JP, Dufresne PJ et al (2015) The *Fasciola hepatica* genome: gene duplication and polymorphism reveals adaptation to the host environment and the capacity for rapid evolution. Genome Biol 16:71
2. Robinson MW, Menon R, Donnelly SM et al (2009) An integrated transcriptomics and proteomics analysis of the secretome of the helminth pathogen *Fasciola hepatica*: proteins associated with invasion and infection of the mammalian host. Mol Cell Proteomics 8:1891–1907
3. Marcilla A, Trelis M, Cortés A et al (2012) Extracellular vesicles from parasitic Helminths contain specific excretory/secretory proteins and are internalized in intestinal host cells. PLoS One 7:e45974
4. Wilson RA, Wright JM, de Castro-Borges W et al (2011) Exploring the *Fasciola hepatica* tegument proteome. Int J Parasitol 41:1347–1359
5. Cwiklinski K, de la Torre-Escudero E, Trelis M et al (2015) The extracellular vesicles of the Helminth pathogen, *Fasciola hepatica*: biogenesis pathways and cargo molecules involved in parasite pathogenesis. Mol Cell Proteomics 14:3258–3273
6. Molina-Hernández V, Mulcahy G, Pérez J et al (2015) *Fasciola hepatica* vaccine: we may not be there yet but we're on the right road. Vet Parasitol 208:101–111
7. Wessel D, Flügge UI (1984) A method for the quantitative recovery of protein in dilute solution in the presence of detergents and lipids. Anal Biochem 138:141–143
8. Pérez-Sánchez R, Valero ML, Ramajo-Hernández A et al (2008) A proteomic approach to the identification of tegumental proteins of male and female *Schistosoma bovis* worms. Mol Biochem Parasitol 161:112–123
9. De la Torre Escudero E, Manzano-Román R, Valero L et al (2011) Comparative proteomic analysis of *Fasciola hepatica* juveniles and *Schistosoma bovis* schistosomula. J Proteome 74:1534–1544
10. de la Torre-Escudero E, Pérez-Sánchez R, Manzano-Román R et al (2013) In vivo intravascular biotinylation of *Schistosoma bovis* adult worms and proteomic analysis of tegumental surface proteins. J Proteome 94:513–526
11. Roberts SM, MacGregor AN, Vojvodic M et al (1983) Tegument surface membranes of adult *Schistosoma mansoni*: development of a method for their isolation. Mol Biochem Parasitol 9:105–127

Chapter 4

Isolation and Analysis of *Fasciola hepatica* Extracellular Vesicles

Alicia Galiano, Maria Teresa Minguez, Christian M. Sánchez, and Antonio Marcilla

Abstract

The finding of extracellular vesicles (EVs) as important players in parasite–parasite and host–parasite communications has led to an increasing number of reports in the literature. Different protocols have been developed for isolation and further characterization of EVs from parasitic helminths. In this chapter, we describe step by step procedures to isolate EVs secreted by *Fasciola hepatica* adults in culture, which could be also applied for other developmental stages of the parasite, as well as EVs present in plasma and urine. Along with classical isolation methods like differential ultracentrifugation, and more recent techniques like size exclusion chromatography (SEC), here we also refer to the storage of EVs for further functional assays. In addition, characterization of *F. hepatica* by electron microscopy techniques like immuno-gold staining, as well as labeling techniques useful for functional assays, like in vitro uptake of fluorescent EVs by cells in culture are also described.

Key words *Fasciola hepatica*, Extracellular vesicles, Isolation, Culture, Analysis

1 Introduction

Extracellular vehicles (EVs) are small membrane-bound organelles that are shed by most cell types. Although once considered to be "cellular garbage cans" with the sole purpose of discarding unwanted cellular material [1], EVs are now recognized as important mediators of intercellular communication by transferring molecular signals, including proteins, mRNA, microRNA, and other noncoding RNA species [2]. Described as exosomes or microvesicles depending on their cellular origin and biogenesis, EVs perform a variety of roles in the maintenance of normal physiology such as blood coagulation, immune regulation, and tissue repair [3], but also participate in pathological settings, notably in tumor progression [3]. EVs have been shown to play important roles in host–parasite, and parasite–parasite communications [4, 5]. Our group first described EVs secreted by parasitic

Martin Cancela and Gabriela Maggioli (eds.), *Fasciola hepatica: Methods and Protocols*, Methods in Molecular Biology, vol. 2137, https://doi.org/10.1007/978-1-0716-0475-5_4, © Springer Science+Business Media, LLC, part of Springer Nature 2020

helminths, in the trematodes *Echinostoma caproni* and *Fasciola hepatica* [6]. Further characterization of *F. hepatica* EVs showed two different types of vesicles with distinct protein cargo [7]. Helminth EVs seem not only to be useful for controlling parasitic diseases but also can modulate the host immune responses. This has been reported using the nematode *Heligmosomoides polygyrus* EVs in mice [8, 9], and very recently we and others have demonstrated that *Nippostrongylus cantonensis* and *F. hepatica* EVs can be used to control inflammatory diseases as ulcerative colitis in a murine model [10, 11].

In this chapter, we present the methodology that can be used to isolate and characterize extracellular vesicles released by *F. hepatica* and from different biological samples.

2 Materials

All buffers are prepared with MilliQ water. Phosphate-buffered saline (PBS) is filtered through 0.22 μm membrane to eliminate contaminants. PBS with a bacteriostatic agent is used for column storage (e.g., 20% ethanol (EtOH) and 0.05% w/v sodium azide).

2.1 Parasite Culture

1. *F. hepatica* adults are obtained from livers collected from infected cattle at local slaughterhouses.
2. RPMI-1640 supplemented with 0.1% glucose.
3. Exosome-depleted fetal bovine serum.
4. PBS: 10 mM sodium phosphate buffer, 150 mM NaCl, pH 7.4.
5. Antibiotics: 100 U penicillin, 100 mg/mL streptomycin.
6. Protease inhibitors cocktail: commercial mix serine and cysteine protease inhibitors.
7. Ethylenediaminetetraacetic acid (EDTA).
8. Tris-buffered saline (TBS): 10 mM Tris–HCl buffer, 150 mM NaCl, pH 7.4.
9. Tris–HCl buffer (TB): 0.05 M Tris–HCl, pH 7.4 (*see* **Note 1**).
10. 50 mL tubes.
11. 10 mL syringes.
12. Aluminum foil.
13. 10% sucrose (sterile filtered).

2.2 Plasma and Urine Sampling

1. Blood and urine are obtained from the infected animals following the recommendations European regulations (Directives 86/609/CEE and 2003/65/CE).

2. 4 mL Vacutainer plastic blood collection tubes with 3.8% buffered sodium citrate.
3. Filter paper.
4. Sodium chloride salt.
5. Membrane filters 0.22 μm.
6. Protein concentration kit.
7. Liquid nitrogen.

2.3 Ultracentrifugation

1. Ultracentrifuge at 120,000 × *g* with compatible fixed angle or swinging bucket rotor.
2. Polycarbonate, polypropylene or similar tubes capable of withstanding forces; 120,000 × *g* and that are compatible with selected rotor and sample input volume.

2.4 Analyses and Characterization of F. hepatica EVs

1. Karnovsky's fixative: 0.1 M sodium phosphate buffer pH 7.2, 2.5% paraformaldehyde, 0.5% glutaraldehyde.
2. 2.5% glutaraldehyde in 0.1 M sodium phosphate buffer pH 7.2.
3. Postfix solution: 0.1 M sodium phosphate buffer pH 7.2, 2% osmium tetroxide.
4. Gold/Palladium (For sample coating).
5. Ethanol (EtOH) 100% to make dilutions (96%, 66%, 33%).
6. LR-white resin.
7. 1% periodic acid in distilled water (dH_2O).
8. 2% sodium metaperiodate in dH_2O.
9. 1% sodium borohydride in dH_2O.
10. 1% ovalbumin in TB.
11. 1% normal goat serum (NGS).
12. Diluted solution: TB, 1% bovine serum albumin (BSA), 0.5% Tween-20.
13. Colloidal gold secondary antibodies (Jackson laboratories).
14. 2% glutaraldehyde in PBS.
15. 2% uranyl acetate.
16. Lead citrate (Reynolds).
17. FM4-64 stain (Molecular Probes Inc.).
18. BODIPY TR Ceramide.
19. SYTO RNA Select.
20. DMSO sterile-filtered.
21. Rat intestinal IEC-19 cells.
22. Human Caco-2 intestinal cells.

2.5 Apparatus

1. Refrigerated centrifuge.
2. Shaker-incubator.
3. Freeze dryer.
4. Microplate absorbance reader.
5. qEV size exclusion columns.
6. Ultracut Leica EM UC6.
7. Transmission electron microscope (Jeol 1010) used at 80 kV with an AMT RX80 (8mpx) digital camera.
8. Scanning electron microscope (Hitachi S4800) with a spotlight of field emission (FEG) with a resolution of 1.4 nm at 1 kV. This equipment has backscattered detector of RX Bruker, transmission detector, the QUANTAX 400 program for micro-analysis and the five motorized axes.
9. Inverted confocal equipment (Olympus FV1000) totally motorized that allows the development of programmed acquisitions on several predefined areas. The lines of excitation are appropriate for experiments in live cells using 405 nm, 488 nm, 515 nm, 559 nm, 594 nm and 635 nm. This instrument also has a CO_2 camera what makes suitable to work with live cells.

3 Methods

3.1 Parasite Culture

1. Collect adult *F. hepatica* parasites (Fig. 1) from cattle livers obtained from a local slaughterhouse.
2. Wash the parasites twice with prewarmed PBS (at 37 °C), and once with prewarmed RPMI1640 culture medium containing 0.1% glucose, 100 U penicillin, and 100 mg/mL streptomycin.

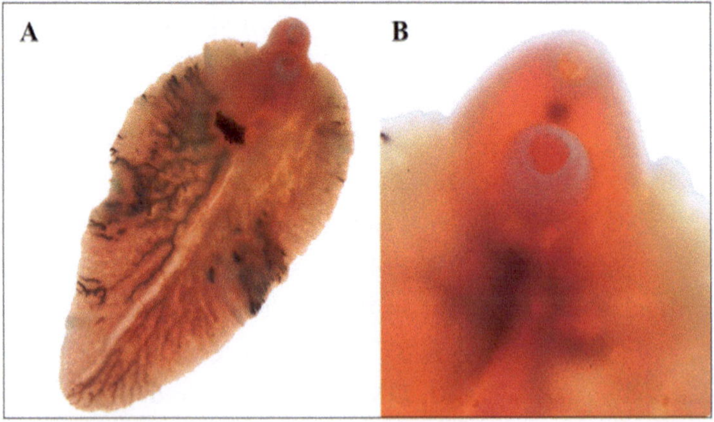

Fig. 1 (**a**) Adult form of *F. hepatica*. (**b**) Detail of the anterior region of the body showing both oral and ventral suckers

3. Incubate 2 adults per 1 mL of RPMI1640 plus penicillin and streptomycin in a 50 mL tube and add proteases inhibitor cocktail. Leave at 37 °C for 3–5 h with low agitation (<100 rpm, checking parasites viability along the experiment).
4. Leave the adults to sediment to the bottom of conical tubes and carefully collect the supernatant after incubation.

3.2 EVs Isolation by Differential Centrifugation

1. Centrifuge the supernatant culture at 700 × *g* for 10 min at 4 °C (pellet contain mainly parasite eggs and debris).
2. Collect the supernatant and distribute it in weighted new 1.5 mL tubes and centrifuge 16,500 × *g* for 20 min at 4 °C (*see* **Note 2**).
3. Collect the supernatants (excretory/secretory products, ES).
4. Filter supernatant through a 0.22 μm membrane filters using 10 mL syringes.
5. Store at 4 °C covered with aluminum foil (prevent from light) if not able to proceed immediately with ultracentrifugation steps.
6. Centrifuge the filtered supernatant at 120,000 × *g* for 60 min at 4 °C.
7. Recover pellet and wash it once with ice cold filtered PBS (1.5 mL).
8. Centrifuge again at 120,000 × *g* for 60 min at 4 °C.
9. Discard supernatant and calculate wet weight of pellets.
10. If using in few hours, EV-pellets can be stored at 4 °C in tubes protected from light, or even gently resuspended with a pipette in 100 μL of filtered PBS. For longer storage, freeze samples at −80 °C in 10% sucrose in filtered PBS.
11. We recommend to get an estimation of the protein concentration using commercial kit (usually it should render 1.0–1.5 μg/μL).

3.3 EVs Isolation from Biological Samples

3.3.1 Initial Processing of Plasma

Plasma samples are obtained from infected cattle from a local slaughterhouse.

Vacutainer plastic blood collection tubes with 3.8% sodium citrate are used to stock up the sample, and preserve at room temperature (RT) until arriving to the laboratory (*see* **Note 3**).

Blood should be gently mixed when is recovered with the sodium citrate buffer to avoid blood clot formation. The processing of the sample in the laboratory is as follows (Fig. 2):

1. Centrifuge the sample at 500 × *g* for 10 min at 4 °C.
2. Collect the supernatant and discard the pellet.
3. Centrifuge the supernatant 2500 × *g* for 20 min at 4 °C.
4. Collect the supernatant and freeze at −80 °C until use.

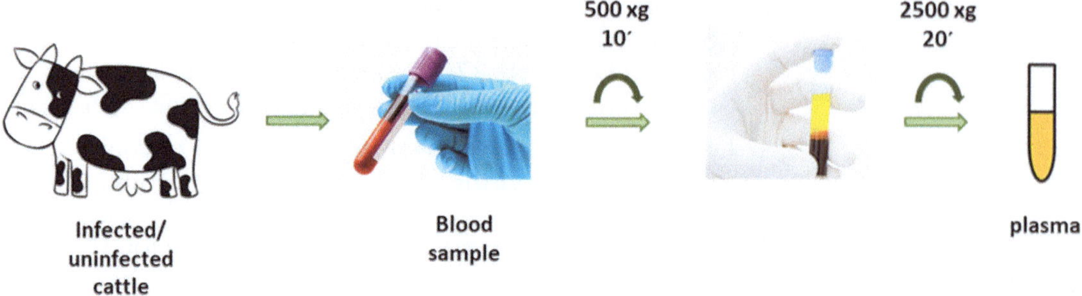

Fig. 2 Experimental procedure followed to isolate EVs from plasma of infected cattle

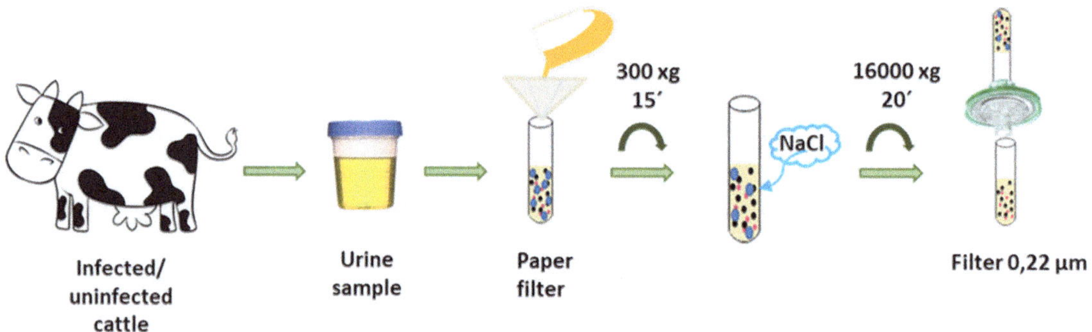

Fig. 3 Experimental procedure followed to isolate EVs from urine of infected cattle

3.3.2 Initial Processing of Urine

Urine samples are collected from infected animals, carried to the laboratory, and processed as described below (Fig. 3):

1. Filter the urine sample in a paper filter to remove impurities.
2. Centrifuge the collected urine at 300 × *g* for 15 min at 4 °C (pellet contain mainly cellular debris).
3. Collect the supernatant and add sodium chloride (NaCl) to a concentration of 0.58 M. Invert containers every 2–3 min). Precipitation should take 2 h at RT (*see* **Note 4**).
4. Centrifuge the urine at 16,000 × *g* for 20 min at 4 °C (it removes remaining cell debris and mucoproteins).
5. Filter supernatant through a 0.22 μm membrane filter using 10 mL syringes (if done properly, it removes vesicles larger than 220 nm).

3.4 EVs Isolation by Size Exclusion Chromatography

1. Prior to the EVs isolation, wash the chromatographic column in filtered PBS (3 vol).
2. Let the PBS reach the stationary phase, and add your sample gently with pipette. Start collecting fractions (0.5 mL) in Eppendorf® tubes in that moment (*see* **Note 5**).

3. Let the added sample reach the stationary phase and start adding filtered PBS (*see* **Note 6**).
4. Collect up to 20 fractions of the eluted sample (if done properly, EVs should be in fractions 7–11) (*see* **Note 7**).
5. Quantify protein concentration in fractions. It usually renders 1.0–1.5 μg/μL in 0.5 mL fractions. Proteins start eluting in later fractions than EVs, and they reach their peak concentration after 3–4 fractions. Notice in which fraction the protein concentration starts to increase; EVs should be in 2–3 fractions previous to those.

3.5 Storage of Functional EVs (Freeze-Drying/Lyophilization in Sucrose 10%)

Before lyophilization measure the protein content of your sample. Place the sample at desired amount and concentration in appropriate vial.

1. Place glass Vacutainer to −80 °C (in order to chill it down to avoid melting of the snap frozen exosomes) (*see* **Note 8**).
2. Add sucrose to the EVs to a final concentration of 10%.
3. Snap-freeze the sample immediately in liquid nitrogen, and keep at −80 °C until lyophilization. Sample must be lyophilized as soon as possible after purification, or anyway within 48 h.
4. Quickly place the tube with the sample in the precooled glass Vacutainer. Be sure that prior to lyophilization you made a small hole in the tube cap (or Parafilm cover).
5. Attach the glass Vacutainer with sample inside the freeze-dryer, and turn on vacuum. Parameters $T = -44$ °C, Pressure = 0.12 mBar (*see* **Note 9**).
6. Take out the sample and seal the cap tightly to avoid humidity.
7. Store the lyophilized exosome samples at 4 °C until reconstitution. Storage at RT should also be efficient.
8. For reconstitution, resuspend the sample(s) in sterile water followed by 1 min pipetting (*see* **Note 10**).

3.6 Analyses and Characterization of F. hepatica EVs

3.6.1 Analysis by Transmission Electron Microscopy (TEM)

For TEM, LR-white resin inclusion was performed as follows:

1. Fix adult parasites and extracellular vesicles covering samples with 2.5% glutaraldehyde in sodium phosphate buffer 0.1 M pH 7.2, or with Karnovsky's fixative solution (2.5% paraformaldehyde; 0.5% glutaraldehyde in sodium phosphate buffer 0.2 M pH 7.2).
2. Wash with phosphate buffer 0.1 M pH 7.2.
3. Postfix with 2% osmium tetroxide in phosphate buffer 0.1 M pH 7.2.

4. Dehydrate samples by sequential treatment in 30% EtOH, 50% EtOH, 70% EtOH and 96% EtOH.
5. Incubate samples in 33% LR-white resin (Sigma) in 96% EtOH for 2 h at RT.
6. Incubate samples in 66% LR-white resin in 96% EtOH for 2 h at RT.
7. Incubate samples in 66% LR-white resin in 100% EtOH for 2 h at RT.
8. Incubate samples in 100% LR-white resin in 100% EtOH for 2 h at RT.
9. Filter samples in resin and polymerize at 60 °C for 48 h.
10. Cut ultrathin slides (60 nm) in an ultracut Leica EM UC6 and stain with 2% uranil acetate.
11. Analyze samples by transmission EM (TEM) using a Jeol JEM1010 microscope at 80 kV.
12. Images are acquired with a digital camera MegaView III with Olympus Image Analysis Software.

3.6.2 Analysis by Scanning Electron Microscopy (SEM) of F. hepatica Adults

For SEM, adult worms are processed as follows:

1. Fix samples covering in Karnovsky's fixative solution (2.5% paraformaldehyde; 0.5% glutaraldehyde in sodium phosphate buffer 0.2 M pH 7.2).
2. Washed once in PBS, 10 min.
3. Postfix for 1–2 h with 2% osmium tetroxide in 0.1 M sodium phosphate buffer pH 7.2.
4. Dehydrate samples by critical point.
5. Mounted specimens are coated with Gold/Palladium, and examined in a scanning electron microscope Hitachi S4800 at 5 kV.

3.6.3 Immuno-Gold Labeling of Fasciola hepatica EVs

Grids are processed in reagents drops as shown in Figs. 4 and 5.

1. Incubate in 1% periodic acid for 10 min.
2. Wash three times with dH_2O (5 min/each).
3. Dry the grids with filter paper.
4. Incubate grids in 2% sodium metaperiodate for 5 min.
5. Wash three times with dH_2O (5 min/each).
6. Dry the grids with filter paper.
7. Incubate grids in 1% sodium borohydride for 10–15 min.
8. Wash five times with dH_2O (5 min/each).
9. Wash once with TB for 3 min.
10. Incubate grids with 1% ovalbumin in TB for 30 min.

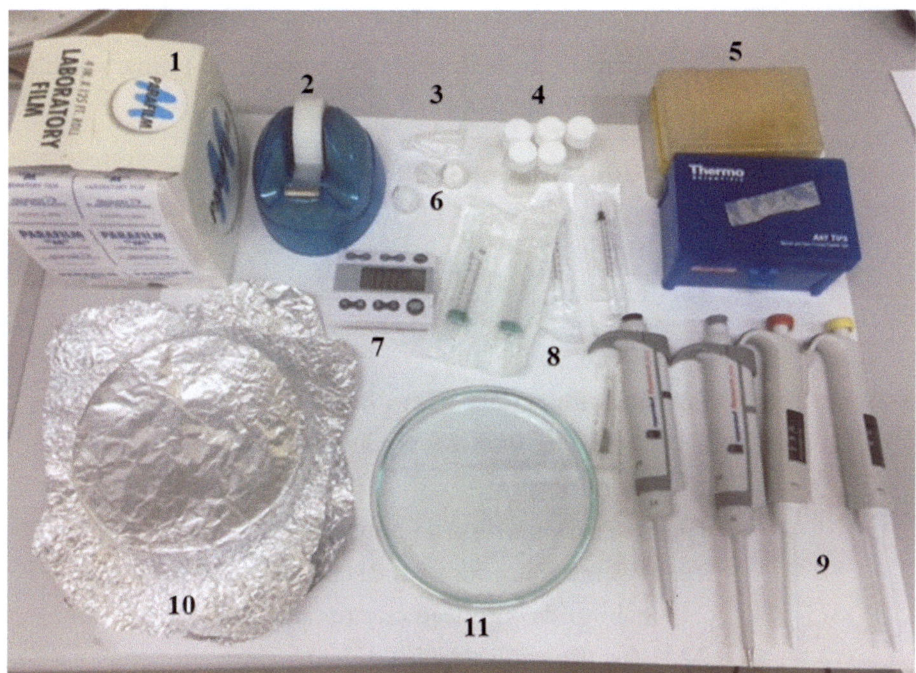

Fig. 4 Materials needed for Immuno-gold staining of EVs. The required materials are: (1) Parafilm® M; (2) Adhesive tape; (3) Eppendorf® tubes; (4) 10 mL containers; (5) Tips; (6) 0.22 μm filters; (7) Timer; (8) Syringes; (9) Automatic pipettes; (10) Aluminum foil; (11) Petri dish

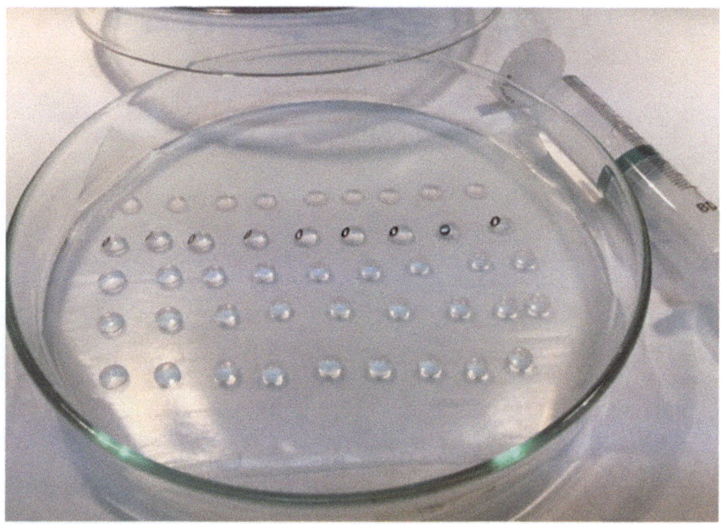

Fig. 5 Nickel grids on different reagents drops in Parafilm® M

11. Wash twice with in TBS (5 min/each).
12. Incubate grids in drop containing the primary antibody solution (usually 1/20 to 1/50) in TBS with 1% NGS for 2 h (*see* **Note 11**).
13. Wash twice with in TBS (5 min/each).
14. Wash once with TB with 1% BSA, plus 0.5% Tween 20 for 10 min.
15. Incubate with colloidal gold secondary antibodies (Jackson laboratories) diluted 1/50 in TB with 1% BSA with 0.5% Tween 20 for 2 h.
16. Wash four times with PBS (7 min/each).
17. Wash once with 2% glutaraldehyde in PBS for 2 min.
18. Wash four times with dH_2O (2 min/each).
19. Stain grids with 2% uranyl acetate for 15 min.
20. Wash three times with dH_2O (2 min/each).
21. Stain grids with lead citrate (Reynolds) for 10 min.
22. Wash thoroughly with dH_2O.
23. View samples by transmission electron microscopy (TEM) at 80 kV.
24. Acquire images with a digital camera. An example is shown in Fig. 6.

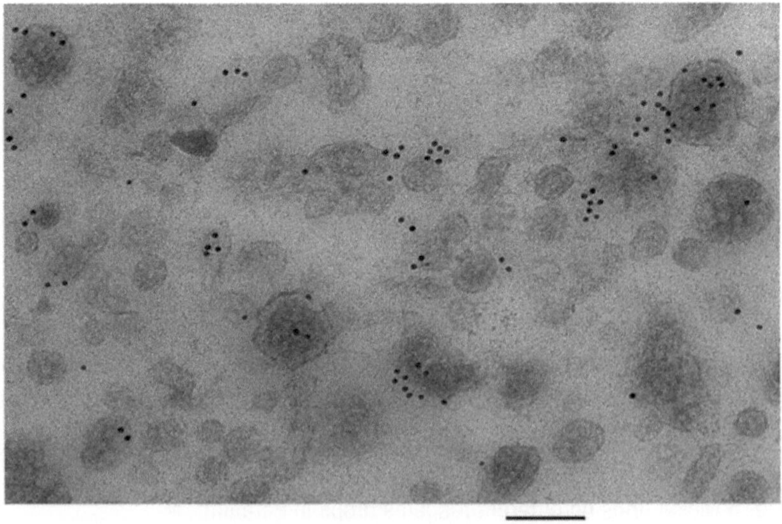

Fig. 6 Immuno-gold staining of *F. hepatica* EVs with anti-FhTP16.5 [12]

3.6.4 Fluorescent Labeling of *Fasciola hepatica* EVs and In Vivo Uptake by Cells in Culture: Confocal Microscopy

Labeling of EVs membranes with FM4-64 stain is done according to the manufacturer's protocol.

1. Purified vesicles are resuspended in 400 μL of PBS, and mixed gently with FM4-64 to a final concentration of 5 μM, for 10 min on ice.
2. Samples are then washed by ultracentrifugation at 120,000 × *g* to remove the excess of dye (*see* **Note 12**).
3. Labeling of vesicles is confirmed by confocal light microscopy in a confocal microscope.
4. Optical excitation is carried out with a 488 nm Argon laser beam and fluorescence emission is detected in the range of 655–698 nm.

3.6.5 Labeling of EVs Membrane and RNA Components

1. Purified EVs are stained with 500 nM of each commercial label: Bodipy TR ceramide and SYTO RNA select (stock solutions of 1 mM solutions were prepared in sterile-filtered DMSO) for 20 min at 37 °C.
2. Samples are washed by Exosome spin columns (MW 3000).
3. Labeling of vesicles is confirmed by confocal light microscopy. Optical excitation is carried out with a 488 nm argon laser beam for RNA Select and with a 594 nm argon laser beam for BODIPY stains, respectively. DAPI is excited with the 405 nm laser beam.

3.6.6 Uptake of Labeled *F. hepatica* EVs

1. Two cell lines have been used, rat intestinal IEC-19 cells and human Caco-2 intestinal cells. A quantity of 40,000 cells/well is incubated in each of μ-Slide 8-well in 200 μL/well of RPMI-1640 with exosome-depleted FBS. At least two slides are used in parallel.
2. Then, approx. 30–40 μL of labeled EVs is added to each well and incubated for 20 min at 37 °C.
3. In the last 10 min of incubation, 25 μL of DAPI (diluted 1/100) is added, trying not to remove cells. Fluorescence uptake by cultured cells is monitored by confocal microscopy at 37 °C (5% CO_2) during a time course of 90–120 min.
4. In parallel, another slide is kept at 4 °C, covered with foil as a control.
5. Slides are examined using a confocal microscopy (Fig. 7), and images are captured continuously using Olympus FluoView (FV Viewer 4.2a) software, at Servicio Central de Soporte a la Investigación Experimental (SCSIE), Universitat de València. Fig. 8 shows a typical result of EVs uptake in Caco-2 cells.

Fig. 7 Olympus FV1000 confocal microscopy with culture chamber

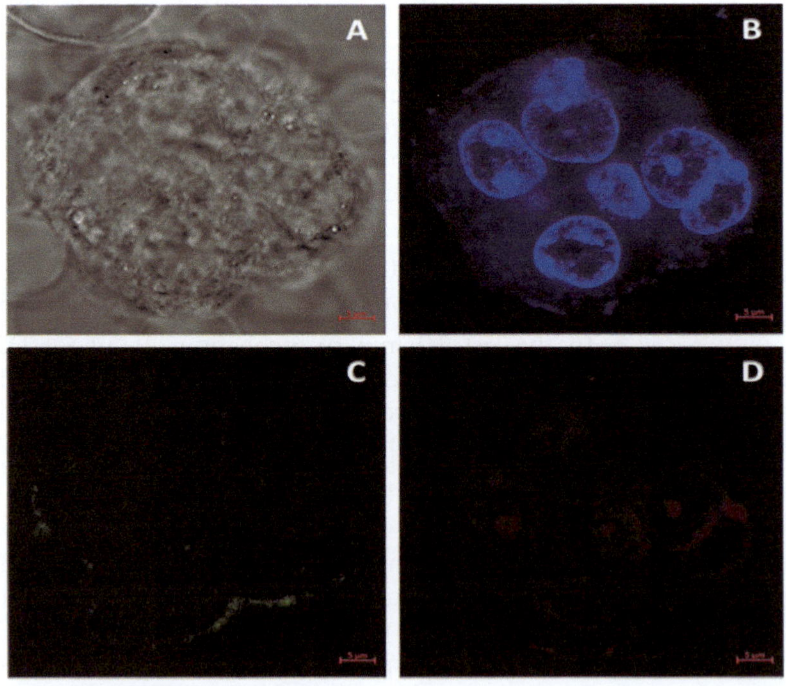

Fig. 8 Labeled EVs from *Fasciola hepatica* are taken up by Caco-2 cells in culture. Panel (**a**): Caco-2 cells under phase contrast; (**b**) DAPI staining showing cells nuclei; (**c**) Green fluorescence corresponding to labeled RNA from *F. hepatica* EVs; (**d**) Red fluorescence corresponding to labeled membranes from *F. hepatica* EVs

4 Notes

1. The reagents are placed in drops on the Parafilm at room temperature, and placed in a Petri dish, and then the grids are changed to the different reagent drops (*see* Fig. 5).
2. Only 1.7 mL Eppendorf tubes are used; do not reuse them.
3. It should not take more than 4 h since the sample is collected until it is processed in the laboratory.
4. This step precipitates urinary mucoproteins like Tamm–Horsfall Protein (THP), helping the following characterization of EVs [13].
5. In order to avoid damaging the column, try to add your sample and PBS slowly and performing a circle movement.
6. The maximum volume you can load onto the column could change depending on the type and the volume of the column, you should not overcome that volume.
7. If the column volume is 10 mL, collect 0.5 mL fractions. If the column volume is 1 mL, collect 0.1 mL fractions.
8. Avoid water in glass Vacutainer.
9. Smaller volumes (up to 230 μL) can be lyophilized in 5 h, but overnight lyophilization is preferred.
10. Exosomes are typically lyophilized in PBS, and the same volume of water is used for reconstitution.
11. Use when primary antibodies are not produced in goat.
12. Alternatively, exosome spin columns can be used, *see* **step2** in Subheading 3.6.5.

Acknowledgments

Dr. Ana M. Espino, Universidad de Puerto Rico, kindly provided rabbit polyclonal antibodies against a *F. hepatica* tegumentary protein 16.5 (anti-FhTP16.5, Gaudier et al., 2012).

The Servei Central de Suport a l'Investigació Experimental (SCSIE) from Universitat de València (Microscopy, Cell Culture and Proteomics) is acknowledged. AM was supported by the Conselleria d'Educació, Cultura i Esports, Generalitat Valenciana, Valencia, Spain (PROMETEO/2016/156), Fundación Ramón Areces, and REDIEX-MINECO.

MTM wishes to express her acknowledgment to Prof. José Manuel Garcia-Verdugo and Prof. Isabel Fariñas for their continuous support.

References

1. Thébaud B, Stewart DJ (2012) Exosomes: cell garbage can, therapeutic carrier, or trojan horse? Circulation 126(22):2553–2555. https://doi.org/10.1161/CIRCULATIONAHA.112.146738
2. El Andaloussi S, Mäger I, Breakefield XO, Wood MJ (2013) Extracellular vesicles: biology and emerging therapeutic opportunities. Nat Rev Drug Discov 12(5):347–357. https://doi.org/10.1038/nrd3978
3. Yáñez-Mó M, Siljander PR, Andreu Z, Zavec AB, Borràs FE, Buzas EI, Buzas K, Casal E, Cappello F, Carvalho J, Colás E, Cordeiro-da Silva A, Fais S, Falcon-Perez JM, Ghobrial IM, Giebel B, Gimona M, Graner M, Gursel I, Gursel M, Heegaard NH, Hendrix A, Kierulf P, Kokubun K, Kosanovic M, Kralj-Iglic V, Krämer-Albers EM, Laitinen S, Lässer C, Lener T, Ligeti E, Linē A, Lipps G, Llorente A, Lötvall J, Mančck-Keber M, Marcilla A, Mittelbrunn M, Nazarenko I, Nolte-'t Hoen EN, Nyman TA, O'Driscoll L, Olivan M, Oliveira C, Pállinger É, Del Portillo HA, Reventós J, Rigau M, Rohde E, Sammar M, Sánchez-Madrid F, Santarém N, Schallmoser K, Ostenfeld MS, Stoorvogel W, Stukelj R, Van der Grein SG, Vasconcelos MH, Wauben MH, De Wever O (2015) Biological properties of extracellular vesicles and their physiological functions. J Extracell Vesicles 4:27066. https://doi.org/10.3402/jev.v4.27066
4. Marcilla A, Martin-Jaular L, Trelis M, de Menezes-Neto A, Osuna A, Bernal D, Fernandez-Becerra C, Almeida IC, Del Portillo HA (2014) Extracellular vesicles in parasitic diseases. J Extracell Vesicles 3:25040. https://doi.org/10.3402/jev.v3.25040
5. Coakley G, Buck AH, Maizels RM (2016) Host parasite communications-messages from helminths for the immune system: parasite communication and cell-cell interactions. Mol Biochem Parasitol 208(1):33–40. https://doi.org/10.1016/j.molbiopara.2016.06.003
6. Marcilla A, Trelis M, Cortés A, Sotillo J, Cantalapiedra F, Minguez MT, Valero ML, Sánchez del Pino MM, Muñoz-Antoli C, Toledo R, Bernal D (2012) Extracellular vesicles from parasitic helminths contain specific excretory/secretory proteins and are internalized in intestinal host cells. PLoS One 7(9): e45974. https://doi.org/10.1371/journal.pone.0045974
7. Cwiklinski K, de la Torre-Escudero E, Trelis M, Bernal D, Dufresne PJ, Brennan GP, O'Neill S, Tort J, Paterson S, Marcilla A, Dalton JP, Robinson MW (2015) The extracellular vesicles of the helminth pathogen, *Fasciola hepatica*: biogenesis pathways and cargo molecules involved in parasite pathogenesis. Mol Cell Proteomics 14(12):3258–3273. https://doi.org/10.1074/mcp.M115.053934
8. Buck AH, Coakley G, Simbari F, McSorley HJ, Quintana JF, Le Bihan T, Kumar S, Abreu-Goodger C, Lear M, Harcus Y, Ceroni A, Babayan SA, Blaxter M, Ivens A, Maizels RM (2014) Exosomes secreted by nematode parasites transfer small RNAs to mammalian cells and modulate innate immunity. Nat Commun 5:5488. https://doi.org/10.1038/ncomms6488
9. Coakley G, McCaskill JL, Borger JG, Simbari F, Robertson E, Millar M, Harcus Y, McSorley HJ, Maizels RM, Buck AH (2017) Extracellular vesicles from a helminth parasite suppress macrophage activation and constitute an effective vaccine for protective immunity. Cell Rep 19(8):1545–1557. https://doi.org/10.1016/j.celrep.2017.05.001
10. Eichenberger RM, Ryan S, Jones L, Buitrago G, Polster R, Montes de Oca M, Zuvelek J, Giacomin PR, Dent LA, Engwerda CR, Field MA, Sotillo J, Loukas A (2018) Hookworm secreted extracellular vesicles interact with host cells and prevent inducible colitis in mice. Front Immunol 9:850. https://doi.org/10.3389/fimmu.2018.00850
11. Roig J, Saiz ML, Galiano A, Trelis M, Cantalapiedra F, Monteagudo C, Giner E, Giner RM, Recio MC, Bernal D, Sánchez-Madrid F, Marcilla A (2018) Extracellular vesicles from the helminth *Fasciola hepatica* prevent DSS-induced acute ulcerative colitis in a T-lymphocyte independent mode. Front Microbiol 9:1036. https://doi.org/10.3389/fmicb.2018.01036
12. Gaudier JF, Cabán-Hernández K, Osuna A, Espino AM (2012) Biochemical characterization and differential expression of a 16.5-kilo Dalton tegument-associated antigen from the liver fluke *Fasciola hepatica*. Clin Vaccine Immunol 19(3):325–333. https://doi.org/10.1128/CVI.05501-11
13. Kosanović M, Janković M (2014) Isolation of urinary extracellular vesicles from Tamm-Horsfall protein-depleted urine and their application in the development of a lectin-exosome-binding assay. BioTechniques 57(3):143–149. https://doi.org/10.2144/000114208

Chapter 5

Cloning and Heterologous Expression of Protein-Coding Sequences in *Escherichia coli*

Martin Cancela and Gabriela Maggioli

Abstract

Recombinant protein expression is widely used to produce large quantities of protein for diverse uses including functional characterization of selected sequences and vaccination trials. In the postgenomic era, high-throughput techniques that allow us to manipulate several sequences are needed. Cloning by in vivo recombination is a technique that consists in the insertion of a linear DNA into a linearized plasmid DNA by in vivo recombination using a recA+ *E. coli* strain. This methodology provides high-throughput cloning with high efficiency without the need for restriction enzyme digestion. In this chapter, we describe two protocols for DNA cloning: one using in vivo recombination and the other by using restriction enzymes. We also describe the application of different conditions to produce functional proteins that needs the incorporation of the amino acid selenocysteine (Sec), like thioredoxin-glutathione reductase enzyme.

Key words *E. coli*, Cloning, Protein expression, In vivo recombination, SECIS element, Selenoprotein

1 Introduction

Postgenomic era demands rapid and high-throughput methods for functional genomic studies [1, 2]. Recombinant protein expression is an easy way to obtain high yields and purity of selected protein for many purposes including functional and structural analysis, screening and profiling of candidate drugs, vaccination studies, and development of diagnostic tools [3]. These studies not only enhance our understanding of molecular aspect of parasite biology and parasitism but also provide novel potential targets for intervention approaches. Bacterial expression systems like *E. coli* are cheap, rapid, and a flexible way to obtain high quantity of active and pure recombinant proteins [1, 4]. Cloning of a protein-coding sequence into plasmid DNA vector is the first step to obtain the desired protein. Different methods allow the introduction of a recombinant DNA into expression vector including the use of restriction enzymes to digest plasmid and gene of interest or by

Martin Cancela and Gabriela Maggioli (eds.), *Fasciola hepatica: Methods and Protocols*, Methods in Molecular Biology, vol. 2137, https://doi.org/10.1007/978-1-0716-0475-5_5,

in vivo recombination [5] that is has as advantages the low cost and avoid the uses of restriction enzyme, preventing undesired gene-sequence cleavage and allowing insertion of any DNA sequence. In this chapter, we describe two protocols for recombinant DNA cloning and protein expression using as examples two protein-encoding sequences: one is an apomucin named *Fh*Muc-1 [6] and the other is a selenoprotein, the thioredoxin glutathione reductase, *Fh*TGR [7]. The selenium is present in diverse organisms as proteins (selenoproteins) in the form of selenocysteine (Sec) which is frequently involved in redox systems. Sec is an amino acid incorporated during protein synthesis and encoded by the UGA codon whose normal function is to terminate translation. Decoding of UGA as Sec involves the Sec insertion machinery and requiring a Sec insertion sequence (SECIS) element in the mRNA, toward 3′ behind this codon [8, 9].

For both recombinant proteins, we use methods previously reported [10–12] with some modifications.

2 Materials

2.1 Medium and Antibiotics

1. LB medium: dissolve 10 g tryptone, 5 g yeast extract, and 10 g NaCl in 1 L dH_2O. Sterilize by autoclaving 15 min.
2. Modified LB medium: dissolve 0.1 g/L cysteine, 0.37 g/L methionine, 2.5 μM sodium selenite in LB medium and supplement with 50 μg/mL kanamycin and 34 μg/mL chloramphenicol [13].
3. 2 YT medium: dissolve 16 g tryptone, 10 g yeast extract, and 5 g NaCl in 1 L of dH_2O. Sterilize by autoclaving.
4. TSB medium: prepare 200 mL of LB liquid medium and adjust pH 6.1. Sterilize by autoclaving and then add 10% (w/v) PEG (MW = 3400), 5% dimethyl sulfoxide, 10% glycerol, 10 mM $MgCl_2$, and 10 mM $MgSO_4$. Sterilize the TSB medium by filtration through 0.22 μm.
5. SOC media: add 20 g tryptone, 5 g yeast extract, 0.58 g NaCl, and 0.18 g KCl to 800 mL of distilled water. Dissolve the mixture by swirling, adjust pH to 7.0 and top up to 1 L. Sterilize by autoclaving and add 10 mL of 1 M $MgCl_2$, 1 M $MgSO_4$, and 1 mL of 2 M glucose. These three solution have to be sterilized by filtration before use.
6. 100 mg/mL kanamycin solution: dissolve 1 g kanamycin in 10 mL Milli-Q water, sterilize by syringe filter of 0.22 μm pore size.
7. 100 mg/mL ampicillin solution: dissolve 1 g ampicillin to 10 mL Milli-Q water, sterilize by syringe filter of 0.22 μm pore size.

8. 34 mg/mL chloramphenicol solution: dissolve 0.34 g of chloramphenicol in 10 mL ethanol 100%, sterilize by syringe filter of 0.22 μm pore size.
9. LB agar plates. To 1 L of LB medium add 15 g Agar. Mix thoroughly and sterilize by autoclaving. Cool the solution to approximately 60 °C and then add the appropriate antibiotic. Then pour the solution onto bacterial culture plates.

2.2 Nucleic Acid Electrophoresis

1. 10× TAE buffer: dissolve 48.4 g of Tris base, 3.72 g disodium EDTA, adjust to pH 8.5 with glacial acetic acid and dilute to 1× TAE solution prior to use.
2. Agarose molecular biology grade.
3. 1% agarose/TAE gel: add 0.40 g of agarose to 40 mL of 1× TAE buffer. Heat until boiling.
4. 0.7% agarose/TAE gel: add 0.28 g of agarose to 40 mL of 1× TAE buffer. Heat until boiling.
5. 5× TBE buffer: dissolve 54 g Tris base, 27.5 g boric acid in distilled water and add 20 mL 0.5 M EDTA pH 8.0. Adjust the final volume to 1 L. Dilute to 1× TBE prior to use.
6. 0.8% agarose/TBE gel: dissolve 0.4 g agarose in 50 mL 1× TBE buffer. Run at 90 V in the same electrophoresis buffer used to prepare agarose gel.
7. Loading dye.
8. Nucleic acid stain for gel electrophoresis: ethidium bromide, Unisafe, GelRed.
9. DNA electrophoresis equipment.
10. Gel documentation system.

2.3 Preparing F. hepatica cDNA

1. *F. hepatica* adult specimen.
2. 0.01 M phosphate-buffered saline (PBS) sterilize by autoclaving.
3. TRIzol Reagent.
4. RNase-free lab plasticware: 50, 15, and 1.5 mL-tubes and pipette tips.
5. Homogenizer.
6. Ice bath.
7. Incubator.
8. Refrigerated centrifuge.
9. RNA extraction kit.
10. Absolute ethanol, molecular biology grade.
11. Diethyl pyrocarbonate (DEPC).

12. 1% agarose/TAE gel in RNase-free conditions (*see* Chapter 7; **steps 8** and **12** in Subheading 2.1).
13. RNase-free DNase I.
14. 25 mM EDTA.
15. cDNA synthesis kit.

2.4 Preparing DNA Vector

1. Plasmid DNA expression vector pGEX-4T3.
2. Restriction enzyme: *Bam*HI, *Eco*RI, or other enzymes that cut at the multiple cloning site.
3. Shrimp alkaline phosphatase (SAP).
4. Incubator (37 °C).
5. DNA electrophoresis equipment to check plasmid DNA cleavage.

2.5 ORF Amplification by PCR

1. *Taq* DNA polymerase high fidelity kit.
2. Gene-specific primers: MucF (5′-CTCCAAAATCGGATCTGGTTCCGCGTGGATCCCCG ATGGTGCGAACGCTAAG-3′) and MucR (5′-CACGATGCGGCCGCTCGAGTCGACCCGGGAATTCC TCAGAAGAACGCAACGC-3′) primers (*see* **Note 1**).
3. 10 mM dNTP mix.
4. Template DNA: *F. hepatica* cDNA or an ORF sequence into a plasmid vector.
5. Thermal cycler for PCR reaction.
6. Check ORF amplification using 0.8% agarose/TBE gel.

2.6 Cloning by In Vivo Recombination Using recA+ E. coli Strain

1. Linearized and dephosphorylated pGEX-4T3 expression vector.
2. ORF amplified by PCR with vector homology sequence.
3. KC8 *E. coli* recA+ strain (to prepare competent cells *see* Subheadings 2.10 and 3.8).
4. 5× KCM: 0.5 M KCl, 150 mM $CaCl_2$, and 250 mM $MgCl_2$. To prepare 50 mL solution: dissolve 1.86 g KCl, 1.1 g $CaCl_2$, and 1.19 g $MgCl_2$ in 50 mL of H_2Odd.
5. LB agar plates supplemented with 50 μg/mL kanamycin and ampicillin (100 μg/mL).
6. Incubator (37 °C).
7. Ice bath.

2.7 Colony PCR to Confirm Recombinant Clones

1. *Taq* DNA polymerase kit.
2. 10 mM dNTP mix.

3. Vector specific primers: pGEX-F (5′-GGGCTGGCAAGCCAC GTTTGGTG-3′) and pGEX-R (5′-CCGGGAGCTGCATGT GTCAGAGG-3′).
4. PCR tubes.
5. Ice bath.
6. Wood stick sterilized by autoclaving.
7. Antibiotics: ampicillin and kanamycin.
8. LB agar plate, 100 μg/mL ampicillin and 50 μg/mL kanamycin.
9. Thermal cycler.
10. LB medium.
11. Incubator.
12. Plasmid DNA extraction kit for pGEX-4T3-FhMuc purification.
13. Check PCR products and DNA plasmid using 0.8% agarose/ TBEgel.

2.8 Production of Recombinant Protein Fused to GST

1. *E. coli* BL21 strain.
2. pGEX-4T3-FhMuc.
3. Electroporation device.
4. Electroporation cuvettes 0.2 mm.
5. 100 mg/mL ampicillin solution.
6. LB agar plates supplemented with 100 μg/mL ampicillin.
7. 2YT medium.
8. Orbital shaker.
9. Spectrophotometer.
10. 0.1 M IPTG: dissolve 0.024 g IPTG in 1 mL dH_2O. Sterilize by filtration using 0.22 μm membrane.
11. Centrifuge.
12. −80 °C freezer.

2.9 Purification of GST-Fused Protein by Glutathione-Sepharose Affinity Chromatography

1. Ultrasonicator.
2. Ice bath.
3. Centrifuge.
4. STE buffer: 10 mM Tris–HCl pH 8.0, 150 mM NaCl, 1 mM EDTA.
5. 30% Sarkosyl.
6. 20% Triton X-100: dilute 20 mL Triton X-100 with PBS in a final volume of 100 mL.

7. Glutathione-Sepharose 4B resin.
8. PBS.
9. Thrombin 1 U/μL.

2.10 Preparation of KC8 Competent Cells

1. KC8 *E. coli* strain.
2. LB agar plates 50 mg/mL kanamycin.
3. 100 mg/mL kanamycin solution.
4. LB medium.
5. TSB.
6. Incubator and shaker.
7. Spectrophotometer.
8. Centrifuge.
9. Ice bath.
10. Liquid nitrogen.
11. −80 °C freezer.

2.11 Cloning FhTGR Wild Type and Their Mutant Sequences into TA Cloning Vector and pET28a Expression Vector

1. *Fasciola hepatica* cDNA.
2. 5TGR: forward primer (5′AGATCTGCTCCAATTCCCGATGATAC-3′).
3. 3TGRSecis: reverse primer (5′-GAATTCATAGGTTAAG CCCCGGTGCAGA CCTGCAACCGGGGCTTAACCTCA GCAAGCGG-3′).
4. 3TGRCys: alternative reverse primers (5′-GAATTCTTAACC ACAGCAAGCG GTCACCTTAGC-3′).
5. 3TGStop: alternative reverse primers (5′-GAATTCTTAACC TCAGCAAGCGG-3′).
6. pCR4-TOPO cloning vector.
7. Appropriate restriction enzymes.
8. Thermal cycler for PCR reaction.
9. Check PCR products and DNA plasmid using 0.7% agarose/TAE gel.
10. Gel extraction kit.
11. T4 DNA ligase and ligation buffer.
12. One Shot® TOP10 Electrocomp™ *E. coli* for cloning.
13. Electroporation device.
14. Shaker and nonshaker incubators.
15. Kanamycin.
16. Chloramphenicol.
17. LB agar plate containing 50 μg/mL kanamycin.

18. LB agar plate containing 50 μg/mL kanamycin and 34 μg/mL chloramphenicol.
19. Quick Plasmid Miniprep Kit.
20. 1× TAE buffer
21. pET28a(+) expression vector.
22. BL21 DE3 *E. coli* strain for recombinant protein expression.
23. pSUABC (*see* **Note 2**).
24. 42 °C water bath.
25. SOC medium.
26. Spectrophotometer to measure optical density of the cell cultures.

2.12 Expression and Purification of Recombinant FhTGR

1. Recombinant *E. coli* (strain BL21DE3) containing FhTGR insert in vector pET28a, in 10% glycerol, store at −80 °C.
2. LB agar plate containing 50 μg/mL kanamycin and 34 μg/mL chloramphenicol.
3. Modified LB medium.
4. 0.1 M IPTG.
5. Sodium selenite.
6. Riboflavin.
7. Niacin.
8. Pyridoxine.
9. Centrifuge to pellet the cells.
10. Lysis Buffer: 50 mM NaH_2PO_4, pH 7.2, 300 mM NaCl, 20 mM imidazole, 1 mg/mL lysozyme.
11. Nickel-nitrilotriacetic acid (Ni-NTA) column: Ni-NTA chelated sepharose (NTA-Sepharose) in a column for Immobilized-metal Affinity Chromatography (IMAC). Add 1 mL of NTA-sepharose into a polypropylene column and add Milli-Q water to wash off the ethanol. Then, add 3–5 vol of 0.4 M nickel sulfate or nickel chloride into the column and wash with Milli-Q water again before equilibrate the column with Lysis Buffer without lysozyme.
12. Wash buffer: 50 mM NaH_2PO_4, pH 7.2, 300 mM NaCl, 30 mM imidazole.
13. Elution buffer: 50 mM NaH_2PO_4, pH 7.2, 300 mM NaCl, 250 mM imidazole.
14. SDS-PAGE analysis.
15. Amicon Ultra-15 Centrifugal Filter, Ultracel-30K.
16. PBS.
17. Protein assay kit.

3 Methods

3.1 Preparing F. hepatica cDNA

1. Adults *F. hepatica* flukes are obtained from the bile ducts of infected cattle from a local abattoir. Liver flukes are extensively washed in PBS at 37 °C. One fluke is used immediately for total RNA extraction.
2. Weight one small worm (~100 mg tissue) in a 1.5 mL tube and homogenize in 1 mL of TRIzol reagent using an ice bath. Thereafter, you can store it at −80 °C for a long time or you can use immediately.
3. Extract total RNA using an RNA extraction kit following the manufacturer's instructions.
4. Confirm the RNA integrity in a 1% agarose/TAE gel using RNase-free conditions and quantify by using 260/280 spectrophotometric measure or RNA quantification kits (e.g., Qubit).
5. Digest contaminant genomic DNA using 2 U of DNase I during 15 min at room temperature. To stop the reaction, add 1 μL of 25 mM EDTA and incubate for 15 min at 65 °C. Then incubate for 1 min on ice bath.
6. 1–5 μg of total RNA (DNase treated) are reverse transcribed using a cDNA synthesis kit following the manufacturer's instructions.
7. cDNA synthesis reaction can be stored at −20 °C or used for PCR immediately. For long periods (>1 year) cDNA should be stored at −80 °C.

3.2 Preparing DNA Vector

1. Digest 1 μg of DNA of selected expression vector (pGEX-4T3) using 2 U of *Bam*HI or other appropriate restriction enzymes (*Eco*RI) in a final volume of 20 μL (*see* **Note 3**).
2. Check 50 ng of digested plasmid DNA in 0.8% agarose-TBE gel.
3. Dephosphorylate 1 μg of plasmid DNA using 1 U of SAP for 30 min at 37 °C. Stop the reaction by incubating for 20 min at 70 °C (*see* **Note 4**).
4. After linearization and dephosphorylation, the vector is suitable for in vivo recombination.

3.3 ORF Amplification by PCR

1. Amplify ORF DNA from cDNA (*see* Subheading 3.1) by using a *Taq* DNA polymerase High Fidelity enzyme and gene-specific primers MucF and MucR with 25 bases sequence homology to pGEX vector (Fig. 1, *see* **Note 5**).
2. Prepare 50 μL of PCR reaction containing 170 μM dNTPs, 1.5 mM $MgSO_4$, 200 μM MucF and MucR primers, and 0.03 U *Taq* DNA polymerase and use 0.5 up to 5 μL of cDNA as template (*see* **Note 6**).

Cloning by *in vivo* recombination in *E. coli*

1. Preparing DNA vector

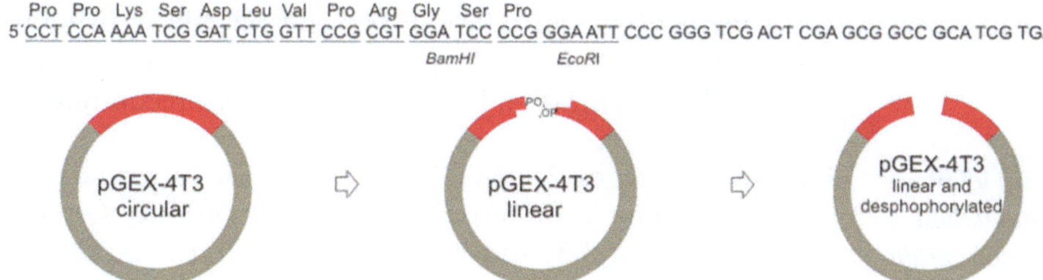

2. ORF amplification by PCR

MucF: 5′-CTCCAAAATCGGATCTGGTTCCGCGTGGATCCCCGATGGTGCGAACGCTAAG-3′
MucR: 5′-CACGATGCGGCCGCTCGAGTCGACCCGGGAATTCCTCAGAAGAACGCAACGC-3′

3. *In vivo* recombination in *E. coli* recA+ strain

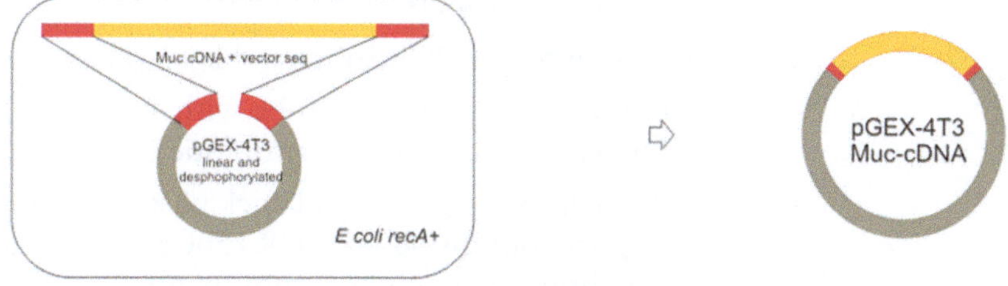

4. Colony PCR to identify recombinant clones

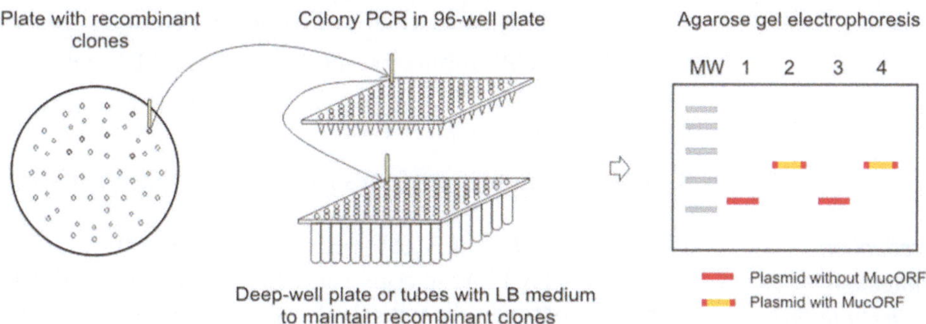

Fig. 1 Overview of the in vivo recombination method using *E. coli* recA+ strain

3. PCR conditions included an initial denaturation at 94 °C for 300 s followed by 5 cycles of 45 s at 94 °C, 60 s at 55 °C, 60 s at 72 °C, 30 cycles 45 s at 94 °C, 60 s at 65 °C, 60 s at 72 °C and a final extension at 72 °C for 10 min (*see* **Note** 7).
4. Amplicons are analyzed by 1.2% agarose gel electrophoresis.
5. Purify the PCR product by using a clean-up kit. Determine DNA concentration before use for in vivo recombination.

3.4 Cloning by In Vivo Recombination in recA+ E. coli Strain

1. Mix 25 μL of 2× KCM with 50 ng of vector (*see* **step 3** in Subheading 3.2) and 50 ng of PCR product (*see* **step 3** in Subheading 3.3) and add water to 50 μL (*see* **Note 8**).
2. Add 50 μL of KC8 *E. coli* cell (*see* **step 3** in Subheading 3.8), mix by inversion and incubate for 20 min on ice.
3. Incubate for 10 min at room temperature.
4. Add 500 μL of LB, mix by inversion and incubate for 3 h at 37 °C.
5. Collect cells by centrifugation for 1 min at 12,000 × *g*. Pour off 900 μL of supernatant, resuspend the pellet cell and plate all the cells on an LB agar plate containing 100 μg/mL of ampicillin.
6. Incubate overnight at 37 °C.

3.5 Colony PCR to Confirm Recombinant Clones

1. Prepare a PCR mix to analyze 10 or more colonies obtained from in vivo recombination experiment. Use a 15–25 μL PCR volume reaction containing 170 μM dNTPs, 1.5 mM $MgCl_2$, 0.1 U/μL *Taq* DNA polymerase, and 300 μM pGEX-F and pGEX-R vector primers.
2. Aliquot in different PCR tubes and keep on ice.
3. Prepare the same number of tube with 1 mL LB 100 μg/mL ampicillin (or the appropriate antibiotic).
4. With the help of a tip or wood stick pick one colony from the LB agar plate, put inside the PCR tube and then inside the LB tube. Mix briefly with circular movements.
5. Perform PCR reactions as follow: an initial denaturation at 94 °C for 5 min and 35 cycles 60 s at 94 °C, 45 s at 50 °C, 60 s at 72 °C and a final extension of 10 min at 72 °C.
6. Analyze PCR amplicons in 0.8% agarose-TBE gel.
7. Culture LB tubes containing the selected clones, overnight at 37 °C with shaking.
8. Select recombinant clones for plasmid DNA purification.

3.6 Production of Recombinant Protein Fused to GST

1. Transform *E. coli* BL21 strain by electroporation using 1 ng of pGEX-4T3-FhMuc construct. Mix 40 μL of *E. coli* BL21 electrocompetent cells with 1 μL (~1 ng pGEX-4T3-FhMuc) into an ice-cold 0.5 mL tube. Put inside a cold 0.2 mm cuvette and

electroporate using an electric pulse of 25 μF capacitance, 2.5 kV and 200 ohm resistance. Add 1 mL SOC medium at room temperature. Incubate 1 h at 37 °C with gentle shaking. Plate different volumes (10, 50, and 200 μL) in LB agar plate with 100 μg/mL of ampicillin. Incubate ON at 37 °C.

2. Pick a single colony and inoculate in 10 mL 2YT 100 μg/mL ampicillin.
3. Incubate ON at 37 °C in a shaking incubator (200 rpm).
4. Transfer 10 mL culture to 1 L 2YT medium with 100 μg/mL ampicillin (*see* **Note 9**).
5. Incubate at 37 °C at 200 rpm until OD600 reaches 0.8–1.
6. Add IPTG to a final concentration of 0.5 mM and incubate for 3 h at 37 °C at 200 rpm.
7. Collect the cells by centrifugation 7000 × *g* 10 min at 4 °C.
8. Discard the supernatant, store the pellet at −80 °C or proceed to purification.

3.7 Purification of GST-Fused Protein by Glutathione-Sepharose Affinity Chromatography

1. Thaw bacterial pellet, and resuspend in 50 mL STE supplemented with 100 μg/mL of lysozyme. Incubate for 15 min on ice bath.
2. Add 1.5% Sarkosyl and 5 mM DTT to the bacterial suspension and ultrasonicate using 5–8 pulses of ultrasound (30 s each, potency 70%) on ice. To avoid sample overheating do a pause of 1 min between each pulse (*see* **Note 10**).
3. Centrifuge at 20,000 × *g*, for 20 min at 4 °C.
4. Collect the supernatant and add Triton X-100 to a final concentration of 2%.
5. Apply 50 mL supernatant onto 1 mL of glutathione-Sepharose 4B resin column preequilibrated in 10 mL PBS.
6. Wash with 100 mL cold PBS.
7. Incubate ON at room temperature with 50 U thrombin in 1 mL PBS.
8. Collect eluted protein.

3.8 Preparation of KC8 Competent Cells

1. Streak KC8 *E. coli* strain in LB agar plate with 20 μg/mL kanamycin and incubate overnight at 37 °C
2. Pick a single colony and inoculate in 2.5 mL LB 20 μg/mL kanamycin.
3. Incubate overnight at 37 °C in a shaking incubator at 180 rpm.
4. Inoculate the 2.5 mL culture to 250 mL LB and incubate until the OD600 is 0.5.
5. Pellet the cells by centrifugation for 15 min at 4 °C at 5000 × *g*.

6. Resuspend the pellet in 12.5 mL ice-cold TSB.
7. Incubate for 10 min on ice.
8. Aliquot the cell (100 μL/tube) and freeze in liquid nitrogen.
9. Store at −80 °C.

3.9 Cloning FhTGR Wild Type and Their Mutant Sequences into TA Cloning Vector and pET28a Expression Vector

1. Amplify by PCR the *Fh*TGR full-length coding sequences using *F. hepatica* cDNA as template and gene-specific primers 5TGR with three different combinations 3TGRSecis, 3TGRCys and 3TGRStop (*see* **Note 11**). PCR conditions included an initial denaturation at 94 °C for 300 s followed by 35 cycles of 30 s at 94 °C, 60 s at 60 °C, 120 s at 72 °C and a final extension at 72 °C for 10 min. Each amplicon was cloned in pCR4-TOPO vector (Fig. 2).
2. Digest 1 μg of plasmids containing the genes of interest with the specific restriction enzymes for cloning into pET28a vector (*see* **Note 12**).
3. Digest 1 μg of pET28a vector with the same restriction enzymes in appropriate 10× buffer.
4. Purify the digested genes of interest and pET28a vector by using gel extraction kit.
5. Ligate the purified digested products and pET28a vector using T4 DNA ligase enzyme and incubate overnight at room temperature.
6. Transform by electroporation the ligated product of recombinant plasmids into TOP10 *E. coli* strain.
7. Spread the mixture on LB agar plate containing 50 μg/mL kanamycin and incubate overnight at 37 °C.
8. Purify each recombinant plasmid from cultured cell by using Plasmid Miniprep Kit.
9. Verify positive recombinant plasmids by DNA sequencing.
10. Transform 5–10 ng of recombinant plasmids into 200 μL *E. coli* BL21(DE3) competent cells previously transformed with pSUABC, by heat shock at 42 °C for 2 min (*see* **Note 13**).
11. Spread the mixture on LB agar plate containing 50 μg/mL kanamycin and 34 μg/mL chloramphenicol and incubate overnight at 37 °C (*see* **Note 14**).

3.10 Expression and Purification of Recombinant FhTGR

1. Inoculate a single positive colony into 10 mL of Modified LB medium in a 50 mL conical tube and incubate at 37 °C with shaking overnight.
2. Inoculate the 5 mL preculture from **step 1** into 500 mL of Modified LB medium in a 500 mL flask and incubate at 25 °C with shaking at 200 rpm until the cultures reach an OD_{600} of 2.4 (*see* **Note 15**).

Cloning using restriction enzymes in *E.coli*

1. ORF amplification by PCR and cloning *Fh*TGRs into pCR4-TOPO

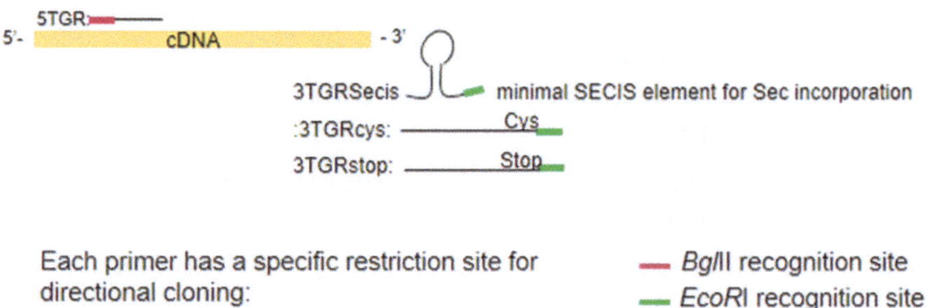

2. *Bgl*II/*EcoR*I digested and ligation with T4 DNA ligase

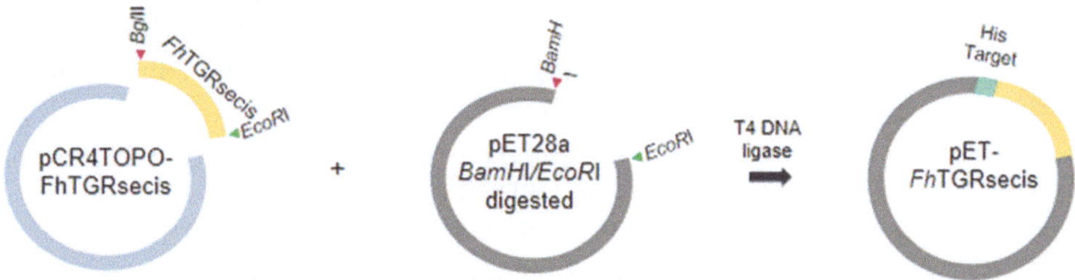

3. Purification of each recombinant plasmid from cultured cell by using Plasmid Miniprep

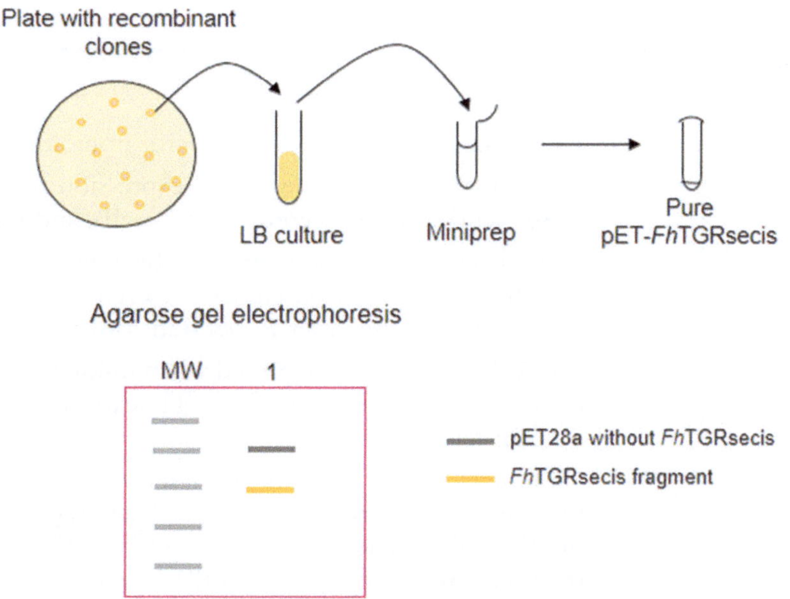

Fig. 2 Overview to construct recombinant plasmid using restriction enzymes. To obtain the three, pET-FhTGRsecis, pET-*Fh*TGRcys, and pET-FhTGRstop vectors, we used the same method. In this figure, we show the step-by-step procedure using pET-FhTGRsecis as an example

3. Induce protein expression with 0.4 mM IPTG.
4. At the same time add 2.5 μM sodium selenite, 20 μg/mL riboflavin, 20 μg/mL niacin, and 20 μg/mL pyridoxine (*see* **Note 16**).
5. Incubate with shaking at 150 rpm for 16 h at 25 °C (*see* **Note 17**).
6. Harvest the cells by centrifugation at 4000 × *g* for 10 min at 4 °C.
7. Suspend the cell pellets in 50 mL Lysis Buffer and incubate for 30 min at room temperature.
8. Sonicate at 10 pulses of 1 min with 1-min pauses.
9. Centrifuge the lysates at 20,000 × *g* for 30 min at 4 °C.
10. Load the supernatant onto the equilibrated Ni-NTA column.
11. Remove the unbound materials from the column by washing with the Wash Buffer.
12. Elute the recombinant proteins with Elution Buffer and analyze them by SDS-PAGE and mass spectrometry (e.g., MALDI-TOF MS/MS).
13. Dialyze 40 mL of eluted protein solution against 2 L of PBS with stirring, overnight at 4 °C.
14. Determine the concentration of the recombinant proteins by Protein assay kit (*see* **Note 18**).
15. Sterilize by filtration and store at −80 °C.

4 Notes

1. Underlined sequences refer to vector homology regions.
2. Plasmid that supports high-level expression of genes involved in Sec synthesis and decoding (selA, selB, and selC) [14].
3. You can alternatively perform a double digestion of DNA using two different restriction enzymes to reduce undesired nonrecombinant colonies after *E. coli* transformation.
4. This step have to be performed when unique digestion enzyme is used in order to avoid ligation of linearized vector after *E. coli* transformation.
5. Cloning by in vivo recombination requires sequence homology regions between vector and ORFs. These regions can be added by PCR using gene-specific primers with additional 25 bases of the vector sequence 5′ to the ORF-specific sequence.
6. In vivo recombination efficiency can be improved using longer vector homology sequences. Longer tags can be added by performing a secondary PCR.

7. To allow annealing of chimeric primer to ORF template, PCR conditions were defined considering an initial Ta that allows chimeric primer to anneal with template. To achieve this calculate Ta of gene-specific region of the primer for the first PCR cycles. After 5 cycles, you can use the Ta of the chimeric primer to specifically amplified your selected sequence.
8. Prepare a tube without PCR product as a negative control. After transformation and plating you can compared number of clones between both positive and negative control. Number of colonies in negative control should be low (<20) to easy detect recombinant clones.
9. In order to achieve a good aeration use 2 L Erlenmeyer with 250 mL of 2YT medium in each one (4 Erlenmeyer for 1 L).
10. To confirm bacterial lysis, you can pipette 1 mL of bacterial suspension (before and after ultrasonication) into a 1.5 mL tube and compare the turbidity of both tubes. You should see clarification of bacterial extract after 5 to 8 pulse of ultrasound.
11. *Fh*TGR is a selenoenzyme with a selenocysteine located at the C-terminal end. In order to assess the relevance of the Sec residue in the recombinant *Fh*TGR produced in *E. coli*, we generated three forms of the protein (wild type, cysteine mutant, and truncated mutant) in pET28a. To obtain the wild type protein, we used the 3TGRSecis primer, which encoded a minimal SECIS element for Sec incorporation, similar to the one present in *E. coli* formate dehydrogenase H [7, 15]. With the 3TGRCys primer produce a Sec/Cys mutation and 3TGRStop contained the native *F. hepatica* TGA codon without an added SECIS element. Each primer has a specific restriction site for cloning into a specific vector, we used the restriction enzymes *BamH*I, *Bgl*II, and *EcoR*I (underlined; *see* **step 2** in Subheading 2.11).
12. Plan this prior to select the suitable vector for recombinant protein expression, we used the restriction enzymes *BamH*I, *Bgl*II, and *EcoR*I.
13. The coexpression with the pSUABC plasmid would improve the enzymatic activity and yields of recombinant TGR. This plasmid is utilized for Sec incorporation during the expression of recombinant selenoproteins in bacterial systems.
14. Antibiotics are used to select the colony containing the two plasmids. The pET28a contains a gene for kanamycin resistance and pSUABC for chloramphenicol resistance.
15. Measure OD_{600} over a 1:10 dilution of the culture.
16. These supplements improve the Sec incorporation.

17. Collect 1 mL of the expression culture at 0, 2, 4, 8, and 16 h to determine the levels of protein expression by SDS-PAGE analysis.
18. Before determining protein concentration, the protein should be concentrated by using Amicon Ultra-15 Centrifugal Filter Units, MWCO 30 kDa. Centrifuge at 4000 × *g* for 30 min at 4 °C.

References

1. Jia B, Jeon CO (2016) High-throughput recombinant protein expression in *Escherichia coli*: current status and future perspectives. Open Biol 6(8):160196
2. Buscaglia CA, Kissinger JC, Agüero F (2015) Neglected tropical diseases in the post-genomic era. Trends Genet 31(10):539–555
3. Fernández-Robledo JA, Vasta GR (2010) Production of recombinant proteins from protozoan parasites. Trends Parasitol 26 (5):244–254
4. Rosano GL, Ceccarelli EA (2014) Recombinant protein expression in *Escherichia coli*: advances and challenges. Front Microbiol 5:172
5. Celie PH, Parret AH, Perrakis A (2016) Recombinant cloning strategies for protein expression. Curr Opin Struct Biol 38:145–154
6. Cancela M et al (2015) Fasciola hepatica mucin-encoding gene: expression, variability and its potential relevance in host-parasite relationship. Parasitology 142(14):1673–1681
7. Maggioli G et al (2011) A recombinant thioredoxin-glutathione reductase from Fasciola hepatica induces a protective response in rabbits. Exp Parasitol 129(4):323–330
8. Bulteau AL, Chavatte L (2015) Update on selenoprotein biosynthesis. Antioxid Redox Signal 23(10):775–794
9. Labunskyy VM, Hatfield DL, Gladyshev VN (2014) Selenoproteins: molecular pathways and physiological roles. Physiol Rev 94 (3):739–777
10. Parrish JR et al (2004) High-throughput cloning of campylobacter jejuni ORfs by in vivo recombination in *Escherichia coli*. J Proteome Res 3(3):582–586
11. Rengby O et al (2004) Assessment of production conditions for efficient use of *Escherichia coli* in high-yield heterologous recombinant selenoprotein synthesis. Appl Environ Microbiol 70(9):5159–5167
12. Bonilla M et al (2008) Platyhelminth mitochondrial and cytosolic redox homeostasis is controlled by a single thioredoxin glutathione reductase and dependent on selenium and glutathione. J Biol Chem 283(26):17898–17907
13. Bar-Noy S, Gorlatov SN, Stadtman TC (2001) Overexpression of wild type and SeCys/Cys mutant of human thioredoxin reductase in *E. coli*: the role of selenocysteine in the catalytic activity. Free Radic Biol Med 30(1):51–61
14. Arnér ES et al (1999) High-level expression in *Escherichia coli* of selenocysteine-containing rat thioredoxin reductase utilizing gene fusions with engineered bacterial-type SECIS elements and co-expression with the selA, selB and selC genes. J Mol Biol 292(5):1003–1016
15. Li C, Reches M, Engelberg-Kulka H (2000) The bulged nucleotide in the *Escherichia coli* minimal selenocysteine insertion sequence participates in interaction with SelB: a genetic approach. J Bacteriol 182(22):6302–6307

Chapter 6

Gene Silencing in the Liver Fluke *Fasciola hepatica*: RNA Interference

Gabriel Rinaldi, Nicolás Dell'Oca, Estela Castillo, and José F. Tort

Abstract

The chronic infection with the liver fluke of the genus *Fasciola* spp. is the most prevalent foodborne trematodiasis, affecting at least one-fourth of the world livestock grazing in areas where the parasite is present. Moreover, fascioliasis is considered a major zoonosis mainly in rural areas of central South America, Northern Africa, and Central Asia. Increasing evidences of resistance against triclabendazole may compromise its use as drug of choice; thus, novel control strategies are desperately needed. Functional genomic approaches play a key role in the validation and characterization of new targets for drug and vaccine development. So far, RNA interference has been the only gene silencing approach successfully employed in liver flukes of the genus *Fasciola* spp. Herein, we describe a detailed step-by-step protocol to perform gene silencing mediated by RNAi in *Fasciola hepatica*.

Key words *F. hepatica*, Gene silencing, RNAi, dsRNA, siRNA, Soaking, Electrosoaking

1 Introduction

Fasciolosis is arguably one of the most widely dispersed zoonotic diseases, affecting livestock worldwide and producing considerable economic losses. Two species of the genus *Fasciola* are responsible for the disease: *Fasciola gigantica* in tropical areas of Africa, Asia, and Oceania, and *Fasciola hepatica* in temperate regions across the globe [1]. Control measures rely mainly on drug treatment, although resistance to the drug of choice, triclabendazole, is emerging in different countries [2]. Studies focused on unraveling the mechanisms of drug resistance are ongoing [3], paralleling long standing efforts to develop vaccines, with variable outputs [4].

The liver fluke *F. hepatica* has entered the postgenomic era when the draft genome was first published in 2015, with a second draft genome from diverse isolates published in 2017 [5, 6]. These

Electronic supplementary material: The online version of this chapter (https://doi.org/10.1007/978-1-0716-0475-5_6) contains supplementary material, which is available to authorized users.

Martin Cancela and Gabriela Maggioli (eds.), *Fasciola hepatica: Methods and Protocols*, Methods in Molecular Biology, vol. 2137, https://doi.org/10.1007/978-1-0716-0475-5_6, © Springer Science+Business Media, LLC, part of Springer Nature 2020

genome data has offered the opportunity to deepen our knowledge of the biology of the parasite and discover novel putative candidates for control strategies. Therefore, functional genomic tools are needed to decipher the function of protein-coding genes and non-coding RNAs [7]. Well ahead of the genome reports, two pioneer reports in 2008 established RNA interference (RNAi)-mediated by double-strand RNA (dsRNA) in *Fasciola* species [8, 9], and this is still now the only reverse genetic tool available so far for these parasitic worms. In order to identify an active RNAi pathway in *F. hepatica*, we first decided to knock down the luciferase mRNA previously delivered into the parasites as an exogenous reporter, followed by RNAi experiments against the *F. hepatica* leucine aminopeptidase [9], a promising vaccine candidate for fasciolosis [10]. At the same time, RNAi was validated as a functional tool by observing a reduced invasion rate of flukes treated by RNAi against cathepsins using an ex vivo assay [8]. Since then, four other reports have extended and optimized the application of this technology in *F. hepatica* [11–13] and more recently in *F. gigantica* [14] (Table 1).

In Schistosome species, RNAi has been successfully applied since 2003 [15, 16] to different developmental stages, including schistosomules, adult worms, eggs. and sporocysts [17]. However, the experience in *Fasciola* spp. has been centered on the newly excysted juvenile (NEJ) stage. This is mainly due to the absence of successful long-term in vitro culture protocols for adult worms, which is critical for RNAi. Moreover, the relatively large size of the adult fluke and the amount of dsRNA needed to reach an optimal concentration in larger culture media make RNAi extremely challenging for the adult stage.

All the reports employed long dsRNA generated by in vitro transcription, since this method is cheap and effective to produce high concentrations of interfering molecules. Since only small interfering RNA (siRNA) can deliver consistent silencing in mammalian cells, this technology has grown, and it has been tested in parallel in all the later RNAi reports in *Fasciola*. Although siRNAs were also effective to induce gene silencing in liver flukes, the level of knockdown was lower and less persistent than the one achieved by long dsRNA molecules [11–13]. In addition, different siRNAs against the same target mRNA resulted in different levels of gene silencing, justifying the use of at least three different siRNA molecules every time a novel target gene is going to be silenced [11]. In *F. gigantica* siRNAs were used only for tracking incorporation of the interfering molecule [14] as was originally employed in *F. hepatica* [9].

Soaking, electroporation, and combined methods have been used as effective delivery strategies for interfering molecules. In our hands, electroporation followed by incubation with the electroporation mixture ("electrosoaking") resulted in the highest

Table 1
Reports on RNAi in *Fasciola* spp.

Species and developmental stage	Targeted genes	RNAi molecule	Delivery	mRNA knockdown	Protein knockdown	Phenotype	References
F. hepatica NEJ	Luc reporter/FhLAP	dsRNA	E	>80% (PCR)	>80% Luc reporter (activity)	ND	[9]
F. hepatica NEJ	FhCL1/FhCB1	dsRNA	S	>80% (PCR)	>75% (immunodetection)	Low motility, (reduced invasion)	[8]
F. hepatica NEJ	FhLAP	dsRNA, siRNA	S, E	>80% (qPCR)	>75% (enzymatic activity)	ND	[11]
F. hepatica NEJ	FhFABP/FhLAP/FhCB/FhCL/ FhGST/FhGST/FhGST/	dsRNA, siRNA	S	>80% (qPCR)	Lagged, (western blot)	None detected	[13]
F. hepatica NEJ	FhCaM1/FhCaM2/FhCaM3	dsRNA, siRNA	S	>80% (qPCR)	>60% (Western blot)	Slow growth Low motility	[12]
F. gigantica NEJ	FgGST	dsRNA	S	>80% (PCR)	ND	ND	[13]
F. gigantica NEJ	FgSOD/FgGST/FgCL1D/ FgCB1-3/FgCB2	dsRNA, siRNA	S, E	>65% (PCR)	|>40% (Western blot)	ND	[14]

NEJ Newly Excysted Juvenile, *E* Electroporation, *S* Soaking, *ND* not determined

incorporation of the silencing molecule and strongest effects in *F. hepatica* [11], with similar results recently reported in *F. gigantica* [14].

The effect of gene silencing can be assessed at different levels. The most obvious is the presence of a detectable phenotype, but its absence does not imply that gene silencing has not been accomplished. It is key to design specific phenotypic and/or behavioral assays based on the expected/predicted gene function of the target, otherwise the results might be unreliable or the phenotype undetectable [18]. The most apparent phenotypes are those related to the organism motility, but specific assays could also be tested (e.g., ex vivo invasion assay) [8]. Care should be taken in interpreting the results, since the latter phenotype (invasive capacity) directly depends on the former (motility).

The presence of phenotypes is not enough to assign the effect to RNAi, and a specific knockdown at the mRNA level should be proved. At the molecular level, the most common method for detecting gene silencing is measuring the levels of mRNA by either end-point RT-PCR or ideally quantitative RT-PCR (RT-qPCR). The reduction of the mRNA levels can also be visualized and localized by in situ hybridization of the mRNA. Effects at the protein level can be analyzed by western blot, enzymatic assays if the target gene encodes an enzyme, or detecting a reduction by immunofluorescent techniques [8].

Critical variables to consider in the design of an RNAi experiment are appropriate controls and number of replicates. Rather than an "untreated" control, in the case of electroporation a "mock" control exposed to the same pulse but with buffer and no dsRNA molecule is more appropriate. Another critical control to study the specificity of the knockdown is an irrelevant dsRNA control (e.g., dsRNA of similar size directed to a target absent in the organism). This control allows the detection of any unspecific effect produced by the presence of a dsRNA molecule regardless of its sequence. Finally, if possible, adding a positive control (i.e., a dsRNA against a target that has already been successfully silenced) might provide more insights into the levels and/or modes of silencing. Ideally, all conditions should be run at least in triplicates to reduce the variability.

Another critical variable to consider in all RNAi experiments is the time to evaluate the effect. The results so far have shown that silencing effects develop after some time and consequently different levels of detection may become evident at different time points depending on the tissue localization of the target gene and half-life of the mRNA and/or protein [13, 18].

Here, we describe a step-by-step protocol to produce consistent silencing in juveniles of *F. hepatica* by electrosoaking, and we comment on the variables and conditions considered in different studies.

2 Materials

2.1 Parasites

1. Metacercariae of *F. hepatica* are obtained from infected *Lymnaea viatrix* snails maintained in the laboratory or purchased from diverse providers.
2. Newly excysted juveniles (NEJs) are obtained by in vitro excystment of metacercariae (*see* Chapter 1; Subheadings 2.6 and 3.6).

2.2 Parasite Culture

1. Inverted microscope.
2. Incubator at 39 °C, 5% CO_2.
3. 6- or 12-well tissue culture plates.
4. 1× Phosphate-buffered saline (PBS)
5. Wash Medium: 500 mL RPMI or DMEM medium, 5 mL 1 M HEPES, 10 mL antibiotic-antimycotic solution (10,000 units/ mL penicillin, 10,000 μg/mL streptomycin, 25 μg/mL amphotericin B).
6. Basch's Medium: to prepare 250 mL add 203.0 mL of 1× Basal medium Eagle, 0.25 g of lactalbumin hydrolysate, 0.25 g of glucose/dextrose, 0.125 mL of 1 mM hypoxanthine, 0.250 mL of 1 mM serotonin, 0.250 mL of 8 mg/mL insulin, 0.250 mL of 1 mM hydrocortisone, 0.250 mL of 0.2 mM triiodothyronine, 1.25 mL of 100× MEM vitamins, 12.5 mL of 1× Schneider's medium, 2.5 mL of 1 M HEPES, 25.0 mL of 1× fetal bovine serum, 5.0 mL of antibiotic–antimycotic solution [19].

2.3 Parasite Viability

1. Stereoscope.
2. Inverted microscope with epifluorescence.
3. Incubator at 39 °C, 5% CO_2.
4. 6- or 12-well tissue culture plates.
5. Pasteur pipettes.
6. Basch's Medium.
7. 1 mM 4′,6-diamidino-2-phenylindole (DAPI) stock solution.
8. 1 mM fluorescein diacetate (FDA) stock solution.
9. 1 mM propidium iodide (PI) stock solution.

2.4 Generation of Interfering and Reporter RNA Molecules

In addition to materials, here we describe the kits which showed best results in our hands (*see* **Note 1**).

1. Thermocycler.
2. Minifuge.
3. Incubators.

4. NanoDrop or spectrophotometer.
5. *Taq* DNA polymerase (*see* **Note 2**).
6. In vitro transcription kit for double strand RNA synthesis (MEGAscript RNAi Kit (Ambion)).
7. 5 M ammonium acetate (NH_4Ac).
8. 100% and 70% ethanol.
9. In vitro transcription kit for mRNA synthesis (mMESSAGE mMACHINE® T7 Ultra Kit- Ambion).
10. RNA purification kit (MEGAclear™ Kit (Ambion)).
11. Silencer® sRNAi Labeling Kit.
12. 5 M NaCl.

2.5 Delivery of Interfering and Reporter RNA Molecules

1. Square-wave electroporator.
2. Gene pulser electroporation cuvettes (4 mm gap).
3. 6 or 12 well tissue culture plates.
4. 1× PBS.
5. Wash Medium.
6. Basch's medium.
7. Interfering and reporter RNA molecules.
8. 4% PFA in 1 × PBS.

2.6 Gene Knockdown Analysis at the mRNA Level

1. Centrifuge.
2. Thermocycler.
3. Real-time PCR equipment.
4. Incubators.
5. RNA extraction kit (NAqueus-Micro Kit (Thermo Fisher AM1931) or the RNeasy Mini kit (Qiagen #74104)).
6. iScript cDNA Synthesis Kit (Bio-Rad, Hercules, CA) or RevertAidTM First Strand cDNA Synthesis Kit (K1621, ThermoFisher).
7. SYBR® Green PCR Master Mix (Thermo Fisher) or the QuantiTect® SYBR® Green PCR (Qiagen).
8. Reference gene primers used in real time PCR experiments:

β-actin	FWD REV	5-GTGTTGGATTCTGGTGATGGTGTC-3 5-CAATTTCTCCTTGATGTCTCG-3
GAPDH	FWD REV	5-GCGCCAATGTTCGTGTTCGG -3 5-TGGCCGTGTACGAATGCAC-3

2.7 Gene Knockdown Analysis at the Protein Level

1. Centrifuge.
2. Luciferase Assay System (Promega, cat. E1500) that includes: 5× Luciferase Cell Culture Lysis Reagents (CCLR), Luciferase Assay Substrate and Luciferase Assay Buffer.
3. Sonicator.
4. Luminometer or Plate Reader (FLUOstar® Omega multi-mode microplate reader).
5. Recombinant luciferase (Promega).
6. Bicinchoninic acid assay (BCA kit, Pierce, Rockford, IL).

2.8 Gene Knockdown Analysis at the Phenotypic Level

1. Stereoscopes.
2. Inverted microscopes with photography and video recording equipment.
3. 0.2% cell culture grade agar in HEPES-buffered Ringer's: 20 mM HEPES pH 7.4, 123 mM NaCl, 5 mM KCl, 1.6 mM $CaCl_2$, 11.1 mM D-glucose.

3 Methods

3.1 *In Vitro* Excystment, Collection and Culture of NEJs

1. Metacercariae (MC) excystment and collection of NEJs was routinely performed as described in Chapter 1, Subheadings 2.6 and 3.6. For our experiments we used both commercially available and *in house* produced metacercariae.
2. Some providers deliver the MC on cellophane films, so they need to be scrapped first. In other cases, they are provided in water without the outer cyst; therefore, we perform a brief hypochlorite wash to clean them (since several algae and bacteria might grow in the batch), and wash thoroughly with 1× PBS prior to activation.
3. After recovering the NEJs, they are washed repeatedly by sedimentation in Wash media to eliminate all traces of excystation medium.
4. Place the NEJs in 12 or 6 well tissue culture plates with either 1–4 mL of Basch's medium respectively and culture at 37 °C, 5% CO_2. For long term culture, change medium every 24–48 h, monitoring the viability of NEJs microscopically (*see* **Note 3**).

3.2 Monitoring the Viability of the Parasites

The viability of parasites can be monitored by double-staining with fluorescent dyes as described for developmental stages of schistosomes [20, 21]. FDA is an esterase substrate used to stain live cells, while PI is only incorporated by damaged cells. Since PI emission overlaps that of Cy3 that we routinely used for following dsRNA uptake, we substituted PI by the nucleic acid stain DAPI as dead cell

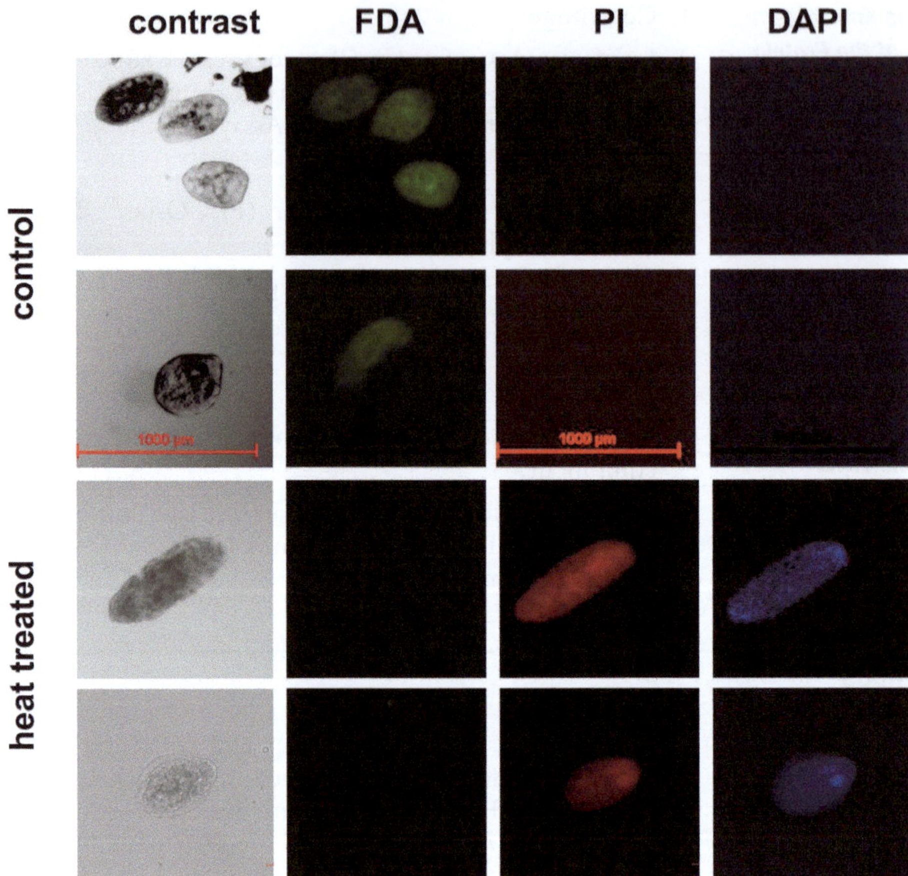

Fig. 1 *F. hepatica* newly excysted juvenile (NEJ) viability assay. NEJs heat-treated for 5 min at 65 °C or live NEJs were incubated for 24 h and costained with FDA, PI or DAPI as indicated Scale bar; 1000 μm

marker [11] (Fig. 1). This protocol allows to count live and damaged worms at different time points.

1. Supplement the Basch's medium with 1 ng/μL DAPI and 0.5 ng/μL FDA and culture normally at 39 °C, 5% CO_2 maintaining the plates covered by aluminum foil.
2. Two hours later, the parasites are washed and transferred to 1× PBS.
3. Monitor the incorporation of the dyes under the fluorescence microscope, using appropriate excitation wavelengths and filters for the respective fluorophores. Parasites are counted with the assistance of ImageJ.

3.3 Design and Preparation of siRNA Molecules

1. siRNA molecules are designed with the assistance of siRNA design tools available on line: IDT siRNA designer tool https://www.idtdna.com/site/order/designtool/index/DSIRNA_CUSTOM) or sBLOCK-iT™ RNAi Designer (https://rnaidesigner.thermofisher.com/rnaiexpress/).

2. The siRNA the Silencer®CyTM3-Labeled NegativeControl #1 s (Ambion, Austin, TX, USA) is employed as the irrelevant control. This sequence is absent from available *F. hepatica* transcriptome data, and the incorporation of Cy3-labeling does not affect the silencing effect (*see* https://www.lifetechnologies.com/order/catalog/product/AM4621).

3.4 Design and Preparation of Long dsRNA Molecules (Longer than 250 bp)

Although both long dsRNA and siRNA molecules triggered a specific gene knockdown, we have previously reported that the gene silencing induced by long dsRNA was stronger and more persistent (e.g., more than 85% gene knockdown was evident by day 21 posttreatment) [11]. We describe a protocol that employs the MEGAscript RNAi Kit which, in our experience, resulted in the highest concentration of dsRNA. The kit was designed to produce high yields of dsRNA longer than 200 bp employing T7 RNA polymerase that transcribes a template DNA flanked by T7 promoter sequences (Fig. 2).

1. Design gene-specific primers with the assistant of a primer design tool and incorporate the T7 promoter region (*TAATACGACTCACTATAGGG*) at the 5′-end of the gene-specific sequence (*see* **Note 4**).
2. Amplify a target gene region of about 500 to 1000 bp using cDNA generated with RNA isolated from NEJ or a plasmid containing the cloned cDNA (*see* **Note 5**).

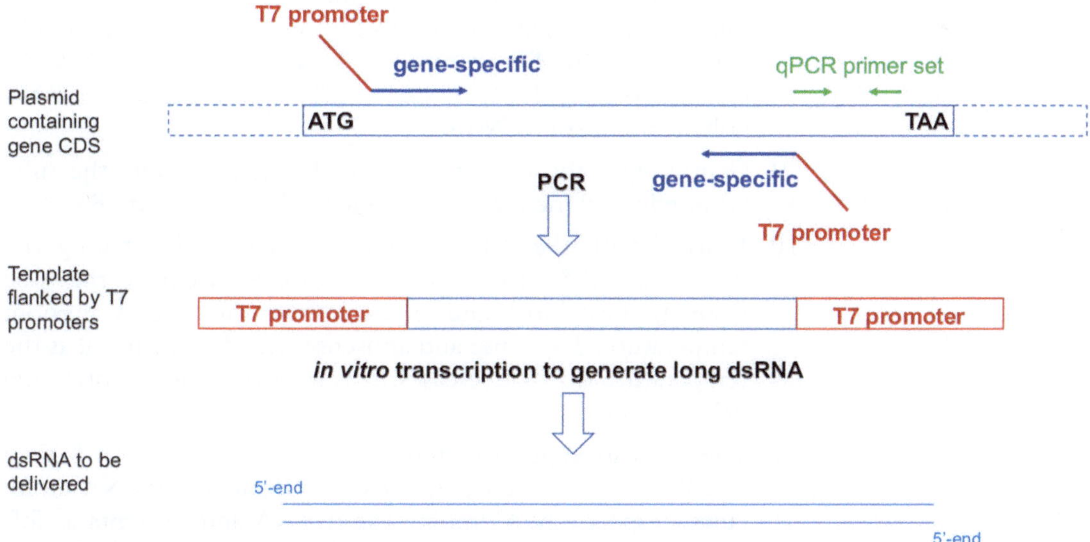

Fig. 2 Generation of long dsRNA by in vitro transcription. Schematic showing the PCR-based approach to generate the target-specific template flanked by T7 promoter sequences for the in vitro transcription reaction and production of long dsRNAs. Primer sets for qPCR reaction indicated in green outside the template region for the dsRNA

3. Amplify the irrelevant dsRNA control (e.g., dsRNA against *Firefly* luciferase (FLuc)) from a commercially available plasmid pGL3-basic plasmid containing the coding sequence for FLuc. This sequence is used as template for PCR using FLuc primers containing T7 promoter sequences at the 5-ends.
4. Set up a PCR reaction to generate the template for the in vitro transcription reaction (*see* **Note 6**). Prepare 50 μL of PCR reaction containing, 46 μL of Platinum™ SuperMix High Fidelity, 200 nM T7 primer forward and T7 primer reverse, and 2 μL of cDNA as template.
5. Perform the PCR with the following general program, an initial denaturation at 94 °C for 3 min followed by 29 cycles of 30 s at 94 °C, 30 s at 50–60∗ °C, 30 s at 68 °C and a final extension at 72 °C for 4 min. ∗Depends on the T_m of the gene-specific sequence of the primer set as recommended [17].
6. Check the product by agarose electrophoresis. Usually, if the PCR product shows a sharp and unique band it is not necessary to purify the PCR product which can be used as template directly in the in vitro transcription reaction.
7. Before proceeding with the in vitro transcription reaction assembly, all the reagents should be thawed. The NTPs and enzyme mix (T7 RNA pol) should be thawed on ice and the 10× T7 RNApol reaction buffer thawed and kept at room temperature.
8. The manufacturer recommends to assembly the reaction in the order shown below and volumes of 20 μL reactions as follows: to 20 μL of nuclease-free water, 5–8 μL of PCR amplicon flanked by T7 promoters, 2 μL of 10× T7 RNA pol reaction buffer, 2 μL of each NTP, and 2 μL of T7 RNA polymerase mix (*see* **Note 7**)·
9. Gently pipet the mixture up and down, centrifuge the tube (s) briefly, and incubate overnight at 37 °C (*see* **Note 8**).
10. Proceed with the dsRNA annealing step by incubating the dsRNA at 75 °C for 5 min (*see* **Note 9**). Remove the tube from the heat block and let cool down the dsRNA at room temperature. The sense and antisense strands will anneal as the temperature slowly decreases. It is important not to place the dsRNA on ice.
11. The nuclease digestion step includes the use of both DNase and RNase to eliminate the DNA template and ssRNA molecules, respectively. Combine the dsRNA and reagents as following, 20 μL of dsRNA, 21 μL of nuclease-free water, 5 μL of 10× digestion buffer, 2 μL of DNase I and 2 μL of RNase.
12. Incubate the reaction at 37 °C for 1 h. It is not recommended to leave the reaction longer than 2 h.

13. The MEGAscript RNAi Kit includes reagents and consumables (tubes and purification columns) for dsRNA purification (*see* **Note 10**). Preheat the Elution Solution at approximately 95 °C. This is suggested for a quicker dsRNA purification protocol.
14. Mix the indicated components as following, 50 μL of dsRNA, 50 μL of 10× binding buffer, 150 μL of nuclease-free water, and 250 μL of 100% ethanol.
15. Mix reaction by gently pipetting up and down and transfer the 500 μL dsRNA mix onto the filter cartridge fitted to a collection tube supplied with the kit.
16. Spin down the tube maximum speed (~ 12,000 × *g*) for 2 min. Discard the flow-through and place the filter cartridge back into the collection tube.
17. Add 500 μL of wash solution onto the filter cartridge, spin down as in the previous step, and discard the flow-through.
18. Repeat previous step for a total of two washes.
19. Discard the flow-through and centrifuge at full speed for 30 s to remove traces of wash solution.
20. Place the filter cartridge onto a new collection tube and apply 50–100 μL (usually 50 μL) of preheated elution solution.
21. Spin down the tube at maximum speed for 2 min.
22. Apply a second aliquot of 50–100 μL (usually 50 μL) of preheated elution solution (**step 11**) and centrifuge as in the previous step. The dsRNA is usually eluted in a total volume of 100 μL.

3.5 dsRNA Quantification and Electrophoretic Analysis

The concentration and purity of the dsRNA can be determined by reading the absorbance in a spectrophotometer at 260 nm (1 A_{260} is equal to 40 μg RNA/mL). We routinely employ NanoDrop microvolume spectrophotometers; however, other protocols to quantify RNA are also recommended (e.g., Qubit RNA assays or Agilent 2100 Bioanalyzer). Check the quality and size of the dsRNA on an agarose gel by running a dilution (1/100 to 1/1000, depending on the original concentration) of the purified dsRNA obtained at the end of **step 22** in Subheading 3.4 (Fig. 3) (*see* **Note 11**).

3.6 Precipitation of the dsRNA (Optional)

When targeting a gene for the first time it is recommended to start with a high concentration of dsRNA (i.e., 30 μg of dsRNA) in the electroporation cuvette if the interfering molecule is delivered by electroporation (see below). However, no more than 30 μL volume of dsRNA in elution solution is recommended because “Wash medium” used as electroporation buffer will be too diluted.

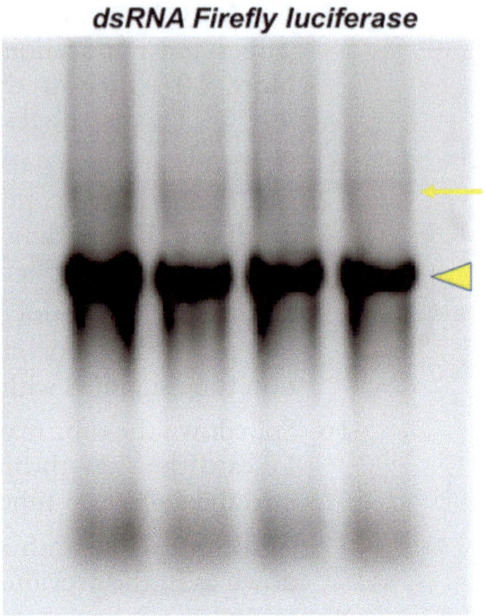

dsRNA Firefly luciferase

Fig. 3 Quality and yield of long dsRNA luciferase molecules synthetized by in vitro transcription. Dilutions of representative Luc dsRNA molecules were run in an agarose gel. The arrowhead indicates the expected size dsRNAs. The arrow indicates higher bands probably due to the presence of dsRNA secondary structures

Therefore, in order to obtain a concentration of 30 μg/μL occasionally would be necessary to precipitate and concentrate the dsRNA as follows,

1. Add 1:10 volume of nuclease-free 5 M NH_4Ac to the purified dsRNA (e.g., if the dsRNA is in 100 μL of elution solution add 10 μL of 5 M NH_4Ac.
2. Add 2.5 volumes of 100% ethanol, 275 μL if the dsRNA was eluted in 100 μL.
3. Mix well by inverting the tube several times, and incubate at −20 °C for at least 30 min. We usually precipitate overnight at −20 °C.
4. Centrifuge at 12,000 × *g* the precipitated dsRNA for 15–30 min at 4 °C and gently remove and discard the supernatant. The pellet of dsRNA is usually evident.
5. Wash the dsRNA pellet with 500 μL of 70% cold ethanol, centrifuge again as above, and remove the supernatant.
6. In order to eliminate the traces of ethanol, centrifuge the dsRNA pellet as above and discard any residual liquid.
7. Air-dry the dsRNA pellet until no more residual liquid is evident and resuspend the pellet using the desired solution and volume. It is convenient to resuspend the dsRNA with the same media used for electroporation protocol, e.g. Wash medium.

8. Quantify the concentrated dsRNA as above and stored at −80 °C until use. It is recommended to store aliquots of the dsRNA to avoid several cycles of freezing and thawing.

3.7 Long dsRNA Labeling

Transfecting parasites with fluorescently labeled interfering RNA molecules allow to track down the molecule incorporation in both live cells and fixed material. In particular, if a transfection protocol is tested for the first time or different conditions of RNAi delivery will be analyzed and compared, it is highly recommended to use labelled interfering molecules [9, 11]. We have used an adapted protocol to label long dsRNA molecules with Cy3 fluorochrome with the assistance of the Silencer® sRNAi Labeling Kit. Importantly, the manufacturer recommends to reduce the exposure to light for the entire procedure.

1. Thaw the kit components and keep on ice. Mix the components as detailed following the indicated order, 20 μL of nuclease-free water, 5 μL of 10× labeling buffer, 20 μL of dsRNA, and 5 μL of labeling dye.
2. Incubate the reaction at 37 °C for 1 h, and proceed with the dsRNA precipitation.
3. To 50 μL labeling reaction add 1:10 volume (5 μL) of nuclease-free 5 M NaCl and 2.5 volume (125 μL) of cold 100% ethanol.
4. Mix thoroughly and incubate at −20 °C for 20–30 min, or overnight.
5. Centrifuge at 12,000 × *g* at 4 °C for 20 min, discard the supernatant, and wash the pellet with 100 μL of 70% ethanol.
6. Centrifuge at 12,000 × *g* for 5 min at room temperature and remove the supernatant.
7. Air-dry pellet and resuspend the labeled dsRNA in 20 μL of nuclease-free water.
8. Store the labeled dsRNA at −80 °C until use.

3.8 Synthesis of Reporter mRNA

To confirm the presence of an active RNAi pathway in organisms for which gene-silencing mediated by dsRNA has not yet been confirmed, we use an exogenous transgene that would be knocked down by a transgene-specific dsRNA [9, 22]. *Fasciola hepatica* NEJs can be transfected with mature *Firefly* luciferase mRNA (mFLuc) as reporter for the presence of an active RNAi pathway [9]. In this section we describe the protocol to generate mature mFLuc, that is, capped mRNA at the 5′-end and polyadenylated at the 3′-end, to be transfected by electroporation into NEJs. It is expected that as soon as the mFLuc is incorporated into the cell cytoplasm it will be translated into protein (*see* **Note 12**).

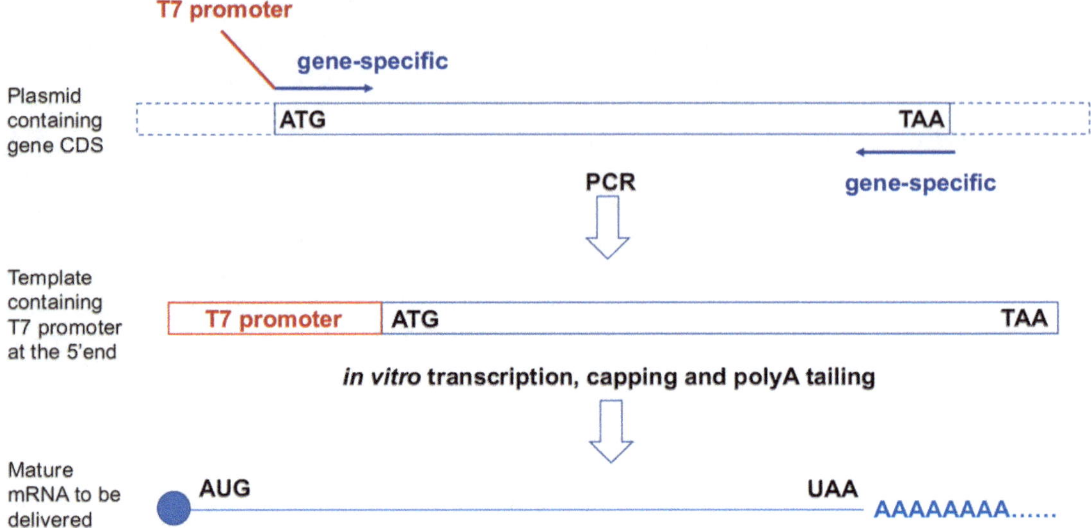

Fig. 4 Generation of 5′-capped, 3′-polyadenylated mature mRNA. Schematic showing the PCR-based approach to generate a reporter template that incorporates the T7 promoter sequence at the 5′-end for the in vitro transcription reaction and production of the 5′-capped, 3′-polyadenylated mature mRNA. It is expected that as soon as the reporter mRNA enters the cytoplasm of the cells it gets translated into protein

1. To generate the template for mRNA synthesis, we performed a PCR reaction on the pGL3-basic plasmid containing the coding sequence for FLuc with a forward primer containing the T7 promoter sequences, following the protocol described in **steps 4** and **5** in Subheading 3.4. This PCR amplicon that contains the T7 promoter sequence at the 5′-end of the gene CDS is used as template for the in vitro transcription reaction (Fig. 4).
2. For in vitro transcription, the synthesis is performed with the mMessage mMachine kit. Before starting thaw all reagents on ice. Vortex the T7 2XNTP/ARCA and 10× T7 RNApol reaction buffer until they are in solution. Keep on ice the T7 2XNTP/ARCA and the T7 RNA polymerase, and store the 10× T7 RNApol reaction buffer at room temperature while the reaction is assembled.
3. Assemble the reaction as recommended by the manufacturer in the order shown below for a 20 μL reaction.
4. To avoid pipetting error and reduce the assembly time a master mix is recommended when several reactions are desired. Mix reaction containing, to 20 μL of nuclease-free water, 10 μL of T7 2XNTP/ARCA, 2 μL of 10× T7 RNApol reaction buffer, 5 μL of PCR amplicon containing T7 promoter at the 5′-end, and 2 μL T7 RNApol mix.
5. Gently pipet the mixture up and down, briefly spin the tube (s) and incubate at 37 °C for 4 h (*see* **Note 13**).

6. At the end of the incubation, briefly spin down the tube to collect the reaction at the bottom of the tube.
7. To eliminate DNA template from the transcription reaction, add 1 μl of DNase to the reaction provided with the kit. The presence of DNA template in the mRNA preparation might affect downstream reactions.
8. Incubate the reaction at 37 °C for 15 min.
9. Add the poly(A) tail at the 3′-end of the capped mRNA by the activity of the poly(A) polymerase E-PAP that employs ATP to elongate the tail at the 3′-end of the transcript (*see* **Note 14**). Assembly the reaction as following, 20 μL of RNA from in vitro transcription (**step 8**), 36 μL of nuclease-free water, 20 μL of 5× E-PAP buffer, 10 μL of 25 mM $MnCl_2$, 10 μL of ATP solution. and 4 μL of E-PAP enzyme.
10. Incubate the reaction at 37 °C for 45 min.

3.9 Purification and Quantification of Mature Capped and Polyadenylated mRNA

Different methods are recommended to recover and purify the 5′-end capped and 3′-end tailed mRNA (*see* **Note 15**). Below we describe the steps for mRNA purification following the MEGAclear Kit protocol. Before starting make sure to add 20 mL of 100% ethanol to the wash solution concentrate and mix well. Usually the mRNA sample will be in 100 μL solution after the poly(A) tailing reaction (**step 9** in Subheading 3.8).

1. Add to the RNA 350 μL of binding solution concentrate and mix gently by pipetting.
2. Add to the sample 250 μL of 100% ethanol to the sample and mix gently by pipetting.
3. Transfer the mRNA mixture onto the filter cartridge fitted into a collection tube supplied in the kit.
4. 15 sec to 1 min at 10,000–15,000 × *g* and discard the flow-through. The collection tube can be reused for the washing steps.
5. Add onto the filter cartridge 500 μL of wash solution and centrifuge as in the previous step.
6. Repeat a second wash with 500 μL of wash solution and centrifuge as in the previous step.
7. Discard the flow-through and continue centrifugation for 10–30 s to eliminate traces of the wash solution.
8. To elute the mRNA, add 50 μL of preheated elution solution (at 95 °C) to the centre of the filter cartridge and spin down the tube for 1 min at room temperature at 10,000–15,000 × *g*.
9. (Optional) Repeat the elution with another 50 μL aliquot of preheated elution solution.

10. (Optional) If required the mRNA can be precipitated and concentrated *see* Subheading 3.6.
11. Quantify the mRNA following the protocol described in Subheading 3.5.

3.10 Delivery of dsRNA by Soaking

1. Prepare a 12- or 24-well culture plate with at least 2 replicate wells for (a) mock control (no dsRNA), (b) irrelevant control (dsRNA of a sequence that is absent in *F. hepatica*) and (c) treated (dsRNA of each target gene).
2. Dilute each dsRNA in Basch's medium at the desired final concentration, we recommend 10 ng/μL for initial study, and aliquot 1 mL per corresponding well (*see* **Note 16**).
3. Aliquot 20–50 NEJs per well and incubate at 37 °C in 5% CO_2 for the desired period, regularly checking the viability of the parasites (*see* **Note 17**). The effect can be detected after 24 h.
4. Every 24 h (or earlier if needed) carefully pipet out most of the media under the stereoscope replacing with 1 mL of prewarmed medium. Store the removed media to detect proteins of interest.

3.11 Delivery of dsRNA by Electrosoaking

1. Calculate the number of conditions to be assayed allowing at least 2 technical replicates per condition.
2. Collect the NEJs after excystment in a 15 mL tube and centrifuge 500 × *g*, 3 min at 4 °C.
3. Remove the media and wash the parasites with wash media three times by centrifugation 500 × *g*, 3 min at 4 °C.
4. Count the parasites in an aliquot to determine the amount of NEJs per condition (>20 NEJs per replicate).
5. Resuspend the parasites in wash medium allowing 100 μL of wash medium per condition to be assayed, and then split the NEJs in equal volumes (100 μL in 1.5 mL microtube) (*see* **Note 18**).
6. Dilute the interfering molecule to 300 ng/μL in wash media calculated for all the replicates.
7. Add 50 μL of each dsRNA solution to the parasites. This brings the final volume to 150 μL (and the final concentration in the solution to 100 ng/μL). Incubate 10 min at RT.
8. Set the electroporator to square wave mode and set the pulse at 125 V, 20 ms (*see* **Note 19**).
9. Transfer the parasites (with the molecules) to a precooled 4 mm gap electroporation cuvette.
10. Place the cuvette in the electroporator and pulse (125 V, 20 ms).

11. Immediately after the pulse, add 850 μL complete prewarmed Basch's medium to the cuvette and carefully mix.
12. Transfer the electroporated parasites from the cuvette to a tissue culture plate. Rinse the cuvette twice with the culture medium to recover most of the worms.
13. Incubate the parasites at 37 °C and 5% CO_2 for the desired time.

3.12 Monitoring RNA Incorporation in Live Parasites

Since the incorporation of Cy3-labeling does not affect the silencing effect it can be used to monitor uptake of the silencing molecules. This can be done by in vitro labeling the dsRNA molecules generated (Subheading 3.7) or using commercial labeled molecules as reporters (*see* **Note 20**).

1. To monitor the dsRNA incorporation in live parasites, collect the parasites and wash them three times with 1× PBS by centrifugation 500 × *g*, 3 min.
2. Washed parasites transfected with Cy3-labeled interfering RNA molecule can be observed under the fluorescence microscope.

3.13 Monitoring RNA Incorporation in Fixed Parasites

Parasites can be fixed and DAPI-stained for a more thorough microscopy observation.

1. Take an aliquot of the treated parasites and wash four times with 1× PBS.
2. Fix in 4% PFA during 30 min at 4 °C.
3. Incubate with 1 μg/mL DAPI for 10 min and wash thoroughly with PBS.
4. Mount on a microscope slide and visualize by fluorescence microscopy (preferably a confocal microscope) at 405 nm and 543 nm to detect DAPI and Cy3, respectively.

3.14 RNA Isolation for Gene Silencing Analysis at the mRNA Level

1. Collect the NEJs in a 1.5 mL microtube.
2. Centrifuge the parasites at 2000 × *g* for 5 min at RT and discard the supernatant.
3. Resuspend the pellet in an appropriate volume of the kit lysis solution according to the provider instructions (usually up to ten volumes of the starting material) (*see* **Note 21**).
4. Disrupt the NEJs using disposable polypropylene RNase-free pellet pestles.
5. Add the appropriate volume of ethanol to the lysate (according to the provider's instructions) in order to precipitate the RNA, and mix by pipetting up and down.
6. Assembly a spin column on the collection tube and transfer the lysate to the spin column and centrifugate at the indicated speed and temperature.

7. Discard the eluate and wash the column with the appropriate volume of wash buffer.
8. Centrifuge to wash, discard eluate and repeat the wash step.
9. Centrifuge 1 min to dry the column.
10. Place the column in a clean recovery tube add carefully 10–20 μL of prewarmed elution solution or RNase-free water. Spin at 8000 × *g* to collect the RNA. If desired repeat the step with a smaller amount of water.
11. Remove any residual DNA by adding one-tenth of volume of 10× DNase buffer and 1 μL of RNase-free DNase I.
12. Incubate 20 min at 37 °C.
13. Inactivate by adding one-tenth of volume of thoroughly mixed DNase inactivation reagent.
14. Mix well and incubate for 5 min at RT.
15. Centrifuge at 10,000 × *g* for 2 min at RT and transfer the supernatant to a new tube. Make sure not to carry over any inactivation resins with the RNA.
16. Measure the RNA concentration and quality by Qubit RNA kit, Agilent 2100 Bioanalyzer or using NanoDrop, or any UV spectrophotometer, at ODs 260 and 280 nm. Reasonable quality RNA has a 260/280 ratio over 1.6.

3.15 Reverse Transcription

Equal amounts of RNA from all conditions (ideally over 50 ng) are used for cDNA synthesis. Different protocols can be used to generate cDNA with equal outcomes. In addition to the method described below, we have successfully used either iScript or RevertAid cDNA synthesis kits (*see* **Note 22**).

1. To eliminate secondary structures in the sample RNA and bind the oligo-dT primers, make a "binding premix" with 1 μL of dNTP mix (10 mM), 1 μL of oligo-dT (0.5 μg/μL), and 5 μL of RNase-free water for one sample. Calculate the amount of reactions (N) to be performed and prepare N+1 samples.
2. Aliquot to "N" different labeled clean PCR tubes 7 μL of the binding solution, and 5 μL of each RNA sample (50–100 ng), and mix carefully by pipetting.
3. Incubate samples at 65 °C for 10 min, and then immediately cool on ice for 3 min.
4. While incubating, prepare an "Enzyme Mix" as follows: 4 μL of 5× RT buffer containing $MgCl_2$ and DTT, 1 μL of RNase inhibitor, 1 μL of reverse transcriptase (200 U), 2 μL RNase-free water. Calculate for N+1 samples.
5. Add 8 μL of Enzyme mix to each RNA–Oligo-dT mixture, mix gently, and spin briefly.

6. Incubate at 42 °C for 1 h.
7. Inactivate the reaction mixtures by heating at 70 °C for 15 min.
8. Chill on ice and spin the reactions and proceed to qPCR or store cDNA at −20 °C.

3.16 Gene Expression Quantitation by Quantitative RT-PCR (RT-qPCR)

Real-time PCR is an efficient way to quantify transcript level variation. In addition to amplifying the target gene, a reference housekeeping gene should be amplified in order to normalize the relative expression of the target gene across the samples [23] (*see* **Note 23**).

1. Prepare a master mix for N+1 reactions to avoid pipetting errors. The typical reaction should be, 1 μL of cDNA sample, 0.5 μL of primer fwd (0.3 μM) and primer rev (0.3 μM), 5 μL of 2× SYBR Master mix, and 3 μL of water.
2. The typical cycling conditions include a preincubation at 95 °C for 10 min to activate the polymerase followed by 35 cycles at 95 °C for 30 s and 60 °C for 30 s, to generate the amplicons. Fluorescence is detected during the annealing/extension step, and melting curve analysis should be performed immediately after PCR cycling. It is expected to have a single peak for each expected amplicon.
3. The relative transcript levels can be estimated using the ΔΔCt method [23]. Statistical analysis is performed by paired Student's t-test on Ct values with the GraphPad Prism Software (www.graphpad.com); p values of ≤ 0.05 were considered significant.

3.17 mRNA Detection by Whole-Mount In Situ Hybridization (WISH)

RNA in situ hybridization in the whole organisms has been extensively used by others and us to qualitatively assess the reduction at the mRNA level (Fig. 5). We refer the reader to a chapter on in situ hybridization protocols applied to *Fasciola hepatica* in the current book (*see* Chapter 7).

3.18 Analysis of the Gene Silencing at the Protein Level

In general, the assays at the protein level would be mainly of three types; the detection of the presence and abundance of the protein on a Western blot, the localization and distribution of the protein by immunohistochemistry, and the quantitation of the enzymatic activity of the protein of interest. Detailed protocols to measure cathepsin L activity are described in Chapter 17, Subheadings 2.1 and 3.2 (*see* **Note 24**). Herein, we describe luciferase as an heterologous reporter [9] and measured the activity (*see* **Note 25**), as follows:

1. At the end of the assay in Subheading 3.11, wash an aliquot of the worms in wash medium and store at −80 °C.
2. Thaw the stored samples and briefly centrifugate at maximum speed 12,000 × *g* to pellet the parasites.

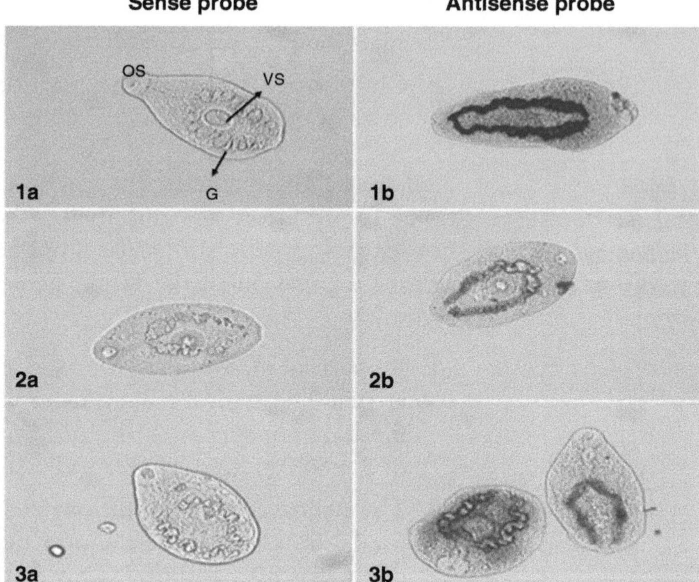

Fig. 5 Detection of cathepsin L2 gene silencing at the mRNA level by RNA in situ hybridization in RNAi-treated NEJs. The NEJs were transfected by soaking with no molecule, that is, mock control (1), dsRNA against cathepsin L2 (2), or a bacterial gene fragment irrelevant dsRNA (3). Thereafter, NEJs were fixed and used in whole-mount in situ hybridization to localize the Cathepsin L2 mRNA using Cathepsin L2 control sense (**a**) or antisense (**b**) probes. *OS* oral sucker, *VS* ventral sucker, *G* gut

3. Remove the medium and briefly resuspend the parasite pellet in 250 μL of 1× CCLR lysis buffer.
4. Sonicate the parasite pellet in 1X CCLR lysis buffer. We routinely use a Heat Systems-Ultrasonics equipment (Plainview, NY, USA) set at 365 s bursts, output cycle 4.
5. Measure the protein concentration of the sonicate with an appropriate protein concentration kit (BCA kit). Recombinant luciferase can be included as a positive control.
6. Aliquots of 100 μL of sonicate are mixed with 100 μL luciferin substrate at RT, and immediately determine the relative light units (RLUs) in the luminometer.
7. Three technical replicates are recommended. Results can be reported as the average of the readings per mg of soluble fluke protein.

3.19 Analysis of the Gene Silencing at the Phenotypic Level

The biological assay employed to detect the phenotype ultimately depends on the expected/predicted function of the target gene. Observation of the gross morphology and motility of the RNAi-transfected parasites is one of the simplest strategies to detect a phenotype. This can be easily followed under the microscope and documented photographically and/or by video (Video S1). McCammick and colleagues [12] developed a simple in vitro assay to measure the parasite migratory capacity that we describe in detail below.

1. Worms are placed into agar substrate at the centre of a petri dish.
2. Place the parasites in the center of the dish where a circle has been drawn and let the parasite migrate freely.
3. Three hours later estimate the proportion of parasites that have migrated beyond a 5 mm radius.
4. Document by photograph and/or by video (Video S1).
5. Count the number of worms displaying normal and aberrant mobility and determine the percentage of affected worms.

4 Notes

1. Labeled or unlabeled long dsRNA molecules (longer than 250 bp) generated in vitro (*see* Subheading 3.3) are routinely used as interfering molecules. In addition, a reporter firefly luciferase mRNA (mLuc) and its corresponding dsRNA can be used to reveal the presence of an active RNAi pathway [9]. Besides long dsRNA, labeled or unlabeled siRNA molecules commercially obtained can be used.
2. Any *Taq* DNA polymerase would be sufficient to produce template for dsRNA generation; however, we prefer the PCR SuperMix High Fidelity (Invitrogen) since its formulation includes a mix of a proofreading enzyme with Taq DNA polymerase that increase the fidelity ~6 times over that of conventional Taq DNA polymerase alone.
3. A very significant increase in parasite survival has been reported by supplementing RPMI with 50% chicken serum [24]. For this reason it appears as an interesting alternative when long-term effects are being studied.
4. In our experience Primer-Blast (https://www.ncbi.nlm.nih.gov/tools/primer-blast/index.cgi?LINK_LOC=BlastHome) has been a successful tool for primer designing.
5. It is recommended that the dsRNA region does not overlap with the PCR-amplicon for the detection of the knockdown (Fig. 2). This could prevent that residual unprocessed long dsRNA that might remain in the RNA preparation from the treated parasite would be reverse-transcribed during the RT reaction and amplified by PCR leading to an underestimation of the gene knockdown [25].

6. During this PCR reaction, the T7 promoter sequences will be incorporated into the amplified fragment generating a template that can be directly transcribed by the T7 RNA polymerase. The forward and reverse T7-primers designed as described above can be employed together in the same PCR reaction generating a template flanked by T7 promoters, or alternatively can be used in separated PCR reactions generating a template for sense and antisense strand transcription by T7 RNA polymerase (Fig. 2). If the PCR template is a plasmid containing the cloned target region, a serial dilution of the plasmid is recommended until the PCR generates a clean and sharp band, (usually a 1/100 or 1/1000 dilution of the original plasmid preparation). cDNA can also be used as template. However, we recommend to clone the template sequence from cDNA and then verify the target DNA by Sanger DNA sequencing. Working with cloned fragments would increase the reproducibility of the assay, eliminating variations due to different cDNA preparations, splicing isoforms or different copies of multigene families that might emerge when working with cDNAs.
7. In our experience, instead of scaling up the reaction volumes, we prefer to increase the number of tubes with 20 μL reaction each per target; for example, typically at least 3 reactions per target are run in parallel in order to obtain enough dsRNA for several RNAi experiments. When running several reactions using several tubes, a master mix (calculating the required amount of reactions plus one) is prepared and a dispensed into several tubes to avoid pipetting errors and to reduce the assembly time. The amount of produced dsRNA depends on the target, for some targets the yield is particularly good, that is, up to ~500 μg of dsRNA per reaction, whereas for other targets and unknown reasons (probably depending on the sequence to be transcribed) the yield is much lower (~100 μg or even less).
8. The manufacturer recommends titrating the incubation time for each target. In our experience, regardless the template the best results have been obtained when overnight incubations were run.
9. The dsRNA annealing step is recommended for transcripts longer than 800 nt generated from a single template flanked by T7 promoters, and for sense and antisense transcripts shorter than 800 nt synthetized in separate transcription reactions. However, we have always included this step regardless the length of the dsRNA.
10. At the end of this step, the dsRNA will be free of proteins, free nucleotides, and nuclei acid digestion products generated in **step 9**.

11. In addition to the expected size band, usually upper bands corresponding to secondary structures or the dsRNA and undigested ssRNA are observed. We have shown these ssRNA bands disappeared with nuclease S1 treatment.

12. The mature mFluc is synthetized with the assistance of the mMESSAGE mMACHINE® T7 Ultra Kit and purified with the MEGAclear™ Kit. To generate a mature mRNA this kit includes Anti-Reverse Cap Analog (ARCA) combined with the other four regular NTPs, (thus, the mixture is called T7 2XNTP/ARCA). In ARCA, one of the 3′ OH groups is substituted with $-OCH_3$ allowing the T7 RNA polymerase to initiate transcription with the remaining –OH group. Therefore, only RNAs capped exclusively in the correct orientation are generated and 100% of the mRNA molecules will be functional.

13. Even though the manufacturer recommends no more than 2 h incubation, we have found the yield increases when the incubation is done for 4 h. Longer incubation might produce lower yield due to mRNA degradation.

14. It is recommended, at least the first time the protocol is run and/ or every time a mRNA of a new reporter is synthetized, to save an aliquot of 2–3 μL of the RNA previous to the polyadenylation step. At the end of the protocol aliquots of both nonpolyadenylated and poly(A) tailed-mRNA can be run in parallel in a gel to check the effective incorporation of the poly (A) tail.

15. Protocols for the purification of the in vitro-transcribed mature mRNA include (1) the use of a kit (Ambion MEGAclear Kit), (2) lithium chloride precipitation, (3) spin column chromatography, and (4) phenol: chloroform extraction followed by isopropanol precipitation. In particular in our hands, the use of the Ambion MEGAclear Kit has provided excellent results by following a quick and simple procedure that removes nucleotides, short oligonucleotides, proteins, and salts from the in vitro-transcribed, capped, and tailed RNA.

16. It is recommended to use the lower concentration that provides a consistent silencing to avoid any off-target effect. We determined that concentrations as low as 2.5 ng/μL allow a consistent silencing by either soaking or electrosoaking, although the silencing effect is more pronounced in the electrosoaking condition [11]. In a similar optimization study McVeigh and collaborators [13] used 50 ng/μL of RNA and diluted it with fresh media after 4 h to 2.5 ng/μL, although they found similarly consistent silencing with 2.5 ng/μL as we reported. The optimization study in *F. gigantica* reported the best results with 50 ng/μL. However, they diluted the starting

media by doubling the volume after 24 h [14]. For this reason, we suggest that for initial testing of an unknown gene by soaking to start with 10 ng/μL. If possible a titration pilot study with concentrations ranging from 2.5 to 50 ng/μL should be performed.

17. This is the smallest amount of NEJs where RNA extraction is consistently viable. Higher number of NEJs per condition, or higher number of replicates would expectedly improve the results.
18. It is important to dilute the parasites in low salt buffers (as wash buffer) rather than full media to avoid the presence of serum and sparking during electroporation that would kill the parasites.
19. The indicated settings are tested in an ECM 830 Square Wave Electroporation System (BTX) https://www.btxonline.com/ecm-830-square-wave-electroporation-system/
20. We used the Silencer® CyTM3-Labeled Negative Control #1 siRNA, both as an irrelevant control in siRNA assays and to track RNA incorporation. This sequence is absent from the *F. hepatica* genome.
21. Several RNA extraction methods and kits can be used. However, we prefer those based on low amount of starting tissue, since this is usually the case. Methods based in column purification are simple and reproducible. In our hands, the RNAqueus-Micro Kit (Thermo Fisher AM1931) or the RNeasy Mini kit (Qiagen #74104) produced good results.
22. We routinely use either the iScript cDNA Synthesis Kit (Bio-Rad, Hercules, CA) or RevertAid™ First Strand cDNA Synthesis Kit (K1621, ThermoFisher) following the manufacturer's instructions with similar results. Prepare the cDNAs obtained from the same assay (diverse treated worms and controls) in parallel, in order to avoid batch effect.
23. Efficiency of each primer set needs to be tested by running serial dilutions of the template and generating a standard curve. The optimal efficiency range for primers amplifying the target gene and housekeeping is 95–105%.
24. An RNAi experiment read out is not complete if the knockdown analysis of the target protein is not performed. The knockdown at the protein level will vary depending on several factors, including the protein half-life and turn over. In addition, an RNAi-associated phenotype is expected when the target protein level has been affected; therefore, RNAi optimization experiments testing the knockdown at different time points after transfection, different dsRNA concentrations, and different delivery regimes are recommended [18].

25. The transfection with the luciferase mRNA (mLuc) is followed by the protein translation within the parasite cells, and the luciferarse activity can be detected using commercial kits (Promega's luciferase assay reagent system) and a Sirius luminometer (Berthold, Pforzheim, Germany).

References

1. Mehmood K, Zhang H, Sabir AJ et al (2017) A review on epidemiology, global prevalence and economical losses of fasciolosis in ruminants. Microb Pathog 109:253–262
2. Kelley JM, Elliott TP, Beddoe T et al (2016) Current threat of Triclabendazole resistance in Fasciola hepatica. Trends Parasitol 32:458–469
3. Hodgkinson J, Cwiklinski K, Beesley NJ et al (2013) Identification of putative markers of triclabendazole resistance by a genome-wide analysis of genetically recombinant Fasciola hepatica. Parasitology 140:1523–1533
4. Beesley NJ, Caminade C, Charlier J et al (2018) Fasciola and fasciolosis in ruminants in Europe: identifying research needs. Transbound Emerg Dis 65(Suppl 1):199–216
5. Cwiklinski K, Dalton JP, Dufresne PJ et al (2015) The *Fasciola hepatica* genome: gene duplication and polymorphism reveals adaptation to the host environment and the capacity for rapid evolution. Genome Biol 16:71
6. Mcnulty SN, Tort JF, Rinaldi G et al (2017) Genomes of Fasciola hepatica from the Americas reveal colonization with Neorickettsia Endobacteria related to the agents of Potomac horse and human Sennetsu fevers. PLoS Genet 13:e1006537
7. International Molecular Helminthology Annotation N, Include ICA, Palevich N et al (2018) Tackling hypotheticals in Helminth genomes. Trends Parasitol 34:179–183
8. Mcgonigle L, Mousley A, Marks NJ et al (2008) The silencing of cysteine proteases in Fasciola hepatica newly excysted juveniles using RNA interference reduces gut penetration. Int J Parasitol 38:149–155
9. Rinaldi G, Morales ME, Cancela M et al (2008) Development of functional genomic tools in trematodes: RNA interference and luciferase reporter gene activity in Fasciola hepatica. PLoS Negl Trop Dis 2:e260
10. Maggioli G, Acosta D, Silveira F et al (2011) The recombinant gut-associated M17 leucine aminopeptidase in combination with different adjuvants confers a high level of protection against Fasciola hepatica infection in sheep. Vaccine 29:9057–9063
11. Dell'oca N, Basika T, Corvo I et al (2014) RNA interference in Fasciola hepatica newly excysted juveniles: long dsRNA induces more persistent silencing than siRNA. Mol Biochem Parasitol 197:28–35
12. Mccammick EM, Mcveigh P, Mccusker P et al (2016) Calmodulin disruption impacts growth and motility in juvenile liver fluke. Parasit Vectors 9:46
13. Mcveigh P, Mccammick EM, Mccusker P et al (2014) RNAi dynamics in juvenile Fasciola spp. liver flukes reveals the persistence of gene silencing *in vitro*. PLoS Negl Trop Dis 8:e3185
14. Anandanarayanan A, Raina OK, Lalrinkima H et al (2017) RNA interference in Fasciola gigantica: establishing and optimization of experimental RNAi in the newly excysted juveniles of the fluke. PLoS Negl Trop Dis 11: e0006109
15. Boyle JP, Wu XJ, Shoemaker CB et al (2003) Using RNA interference to manipulate endogenous gene expression in Schistosoma mansoni sporocysts. Mol Biochem Parasitol 128:205–215
16. Skelly PJ, Da'dara A, Harn DA (2003) Suppression of cathepsin B expression in Schistosoma mansoni by RNA interference. Int J Parasitol 33:363–369
17. Da'dara AA, Skelly PJ (2015) Gene suppression in schistosomes using RNAi. Methods Mol Biol 1201:143–164
18. Dalzell JJ, Warnock ND, Mcveigh P et al (2012) Considering RNAi experimental design in parasitic helminths. Parasitology 139:589–604
19. Basch PF (1981) Cultivation of Schistosoma mansoni *in vitro*. I. Establishment of cultures from cercariae and development until pairing. J Parasitol 67:179–185
20. Peak E, Chalmers IW, Hoffmann KF (2010) Development and validation of a quantitative, high-throughput, fluorescent-based bioassay to detect schistosoma viability. PLoS Negl Trop Dis 4:e759
21. Yan HB, Smout MJ, Ju C et al (2018) Developmental sensitivity in Schistosoma mansoni to Puromycin to establish drug selection of

transgenic Schistosomes. Antimicrob Agents Chemother 62:8

22. Rinaldi G, Okatcha TI, Popratiloff A et al (2011) Genetic manipulation of Schistosoma haematobium, the neglected schistosome. PLoS Negl Trop Dis 5:e1348
23. Livak KJ, Schmittgen TD (2001) Analysis of relative gene expression data using real-time quantitative PCR and the 2(−Delta Delta C (T)) method. Methods 25:402–408
24. Mccusker P, Mcveigh P, Rathinasamy V et al (2016) Stimulating Neoblast-like cell proliferation in juvenile Fasciola hepatica supports growth and progression towards the adult phenotype *in vitro*. PLoS Negl Trop Dis 10: e0004994
25. Morales ME, Rinaldi G, Gobert GN et al (2008) RNA interference of Schistosoma mansoni cathepsin D, the apical enzyme of the hemoglobin proteolysis cascade. Mol Biochem Parasitol 157:160–168

Chapter 7

Analysis of Gene Expression in *Fasciola hepatica* Juveniles and Adults by In Situ Hybridization

Estela Castillo and Uriel Koziol

Abstract

In situ hybridization (ISH) is a technique used for the spatial localization of nucleic acids within tissues and cells. It is based on the ability of labeled nucleic acids (probes) to hybridize under the right conditions with the nucleic acids present in fixed biological specimens. In this chapter, we describe protocols for detection of RNA by ISH using digoxigenin (DIG)-labeled probes for *Fasciola hepatica* adults (in cryosections, given their large size) and for newly excysted juveniles (NEJs, which are ideally suited given their small size for whole-mount ISH). We describe fluorogenic and chromogenic protocols, respectively, but the detection methods can be easily interchanged by using the appropriate enzyme-conjugated antibodies and detection solutions.

Key words *Fasciola*, In situ hybridization, WMISH, FISH, Gene expression

1 Introduction

In situ hybridization (ISH) is a technique used for the spatial localization of nucleic acids within tissues and cells. It is based on the ability of labeled nucleic acids (probes) to hybridize under the right conditions with the nucleic acids present in fixed biological specimens. It can be used for the detection of DNA or RNA molecules, depending of the specific conditions used in the experiment. Although originally developed with radioactively labeled probes, nowadays the detection of the labeled probe is in most cases indirect, using an enzyme-conjugated antibody that can detect a nonradioactive label present in the probe [1]. The enzymes used for detection are usually alkaline phosphatase or horseradish peroxidase. Depending on which substrate is used for the development of the signal, the reaction product may be colored (for chromogenic in situ hybridization) or fluorescent (for fluorescent in situ hybridization).

ISH can be performed in tissue sections of large specimens, but can also be performed in whole mounts (WMISH) in the case of

Martin Cancela and Gabriela Maggioli (eds.), *Fasciola hepatica: Methods and Protocols*, Methods in Molecular Biology, vol. 2137, https://doi.org/10.1007/978-1-0716-0475-5_7,

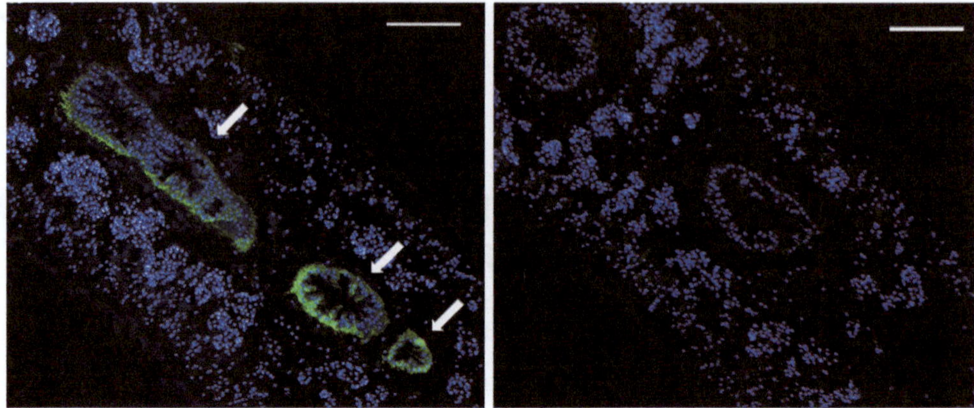

Fig. 1 Fluorescent ISH on sagittal sections of *F. hepatica* adults using a *cathepsin L2* probe (left panel) and a control probe (right panel). Notice the strong specific signal in the cross sections of the gut branches in the left panel. Bars represent approximately 100 μm

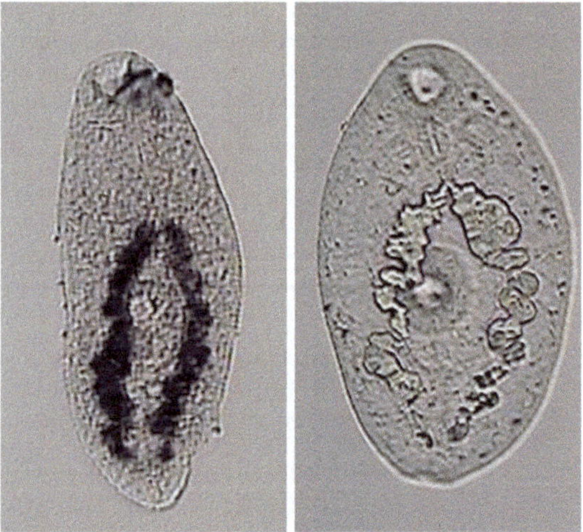

Fig. 2 Chromogenic WMISH of *F. hepatica* NEJ using a *cathepsin L2* probe (left panel) and a control probe (right panel). Notice the strong specific signal in the gut branches in the left panel

small specimens such as embryos and small invertebrates. In this chapter, we describe protocols for detection of RNA molecules by ISH using digoxigenin (DIG) labeled probes for *F. hepatica* adults (in cryosections, given their large size) and for newly excysted juveniles (NEJs, which are ideally suited given their small size for WMISH). We describe fluorogenic and chromogenic protocols, respectively, but the detection methods can be easily interchanged by using the appropriate enzyme-conjugated antibodies and detection solutions. We provide examples of results from both protocols using a probe against *cathepsin L2* which is expressed in the digestive system of adults and NEJs of *F. hepatica* respectively ([2] and unpublished results) (Figs. 1 and 2).

2 Materials

2.1 Materials for In Vitro Labeling of RNA Probes with DIG

1. Nuclease-free ultrapure water (*see* **Note 1**).
2. DIG RNA labeling solution, containing unlabeled NTPs and digoxigenin-UTP (Merck 11277073910).
3. Purified cDNA fragment of the gene to be analyzed, with different viral RNA polymerase promoters (e.g., T7, SP6, T3 promoters) located upstream and downstream of the fragment (*see* **Note 2**).
4. Viral RNA polymerases compatible with the promoters found upstream and downstream of the cDNA fragment (e.g., T7, SP6, T3 RNA polymerase). These are provided by commercial suppliers with their appropriate reaction buffers.
5. Recombinant RNase inhibitors (e.g., RNaseOUT, RiboLock, RNasin).
6. Water bath.
7. DNase (RNase free).
8. DEPC-treated water: Treat double-distilled water with 0.1% diethyl pyrocarbonate (DEPC) overnight with shaking, and autoclave for 40 min.
9. 3 M Sodium acetate solution pH 5.2: Prepare in DEPC-treated water overnight with shaking, and autoclave for 40 min.
10. Absolute ethanol, molecular biology grade.
11. Agarose, molecular biology grade.
12. TAE Buffer: 40 mM Tris, 20 mM acetic acid, 1 mM EDTA, dissolved in DEPC-treated double-distilled water.

2.2 Materials for Probe Quantification by Dot-Blot

1. DIG-labeled control DNA (Merck 11585746910).
2. Positively charged nylon membrane.
3. Plastic container.
4. Whatman filter paper.
5. UV transilluminator.
6. Dot-Blot washing buffer: 100 mM Tris–HCl pH 7.5, 150 mM NaCl.
7. Dot-Blot blocking buffer: 100 mM Tris–HCl pH 7.5, 150 mM NaCl. Add 0.5% (w/v) Blocking reagent (Merck 11096176001) and dissolve by heating at 65 °C with periodic shaking.
8. Anti-digoxigenin antibody conjugated to alkaline phosphatase (Anti-DIG-AP, Merck 11093274910).
9. Alkaline phosphatase buffer: 100 mM Tris–HCl pH 9.5, 100 mM NaCl, 50 mM $MgCl_2$.

10. Nitro-blue tetrazolium and 5-bromo-4-chloro-3′-indolyphosphate (NBT/BCIP solution). Dissolve 1 g NBT in 20 mL of 70% dimethylformamide (DMF). Dissolve 1 g BCIP in 20 mL of 100% DMF. Add 33 μL of BCIP and 66 μL of NBT per 10 mL of alkaline phosphatase buffer just before adding to membrane.

2.3 Materials for Fluorescent ISH on Sections of F. hepatica Adults

1. DEPC-treated water.
2. Phosphate Buffered Saline (PBS)-DEPC, (PBS-DEPC): 137 mM NaCl, 2.7 mM KCl, 10 mM Na_2HPO_4, 2 mM KH_2PO_4. Adjust pH to 7.4 with HCl. Treat with 0.1% DEPC overnight with shaking, and autoclave for 40 min.
3. Paraformaldehyde/PBS fixation solution (PFA/PBS-DEPC): Dissolve 4 g of PFA in 100 mL of DEPC-treated PBS by incubating at 60 to 70 °C for 1–2 h and shaking periodically. Make 10 mL aliquots and store at −20 °C.
4. Sucrose solution: 30% of molecular biology grade sucrose in DEPC-treated double distilled water.
5. OCT compound mounting medium.
6. Aluminum foil. Eliminate RNases with dry heat (180 °C for 2 h).
7. Liquid nitrogen.
8. Plastic container.
9. Cryostat.
10. Microscope slides, RNase free, coated.
11. Coplin Jars. Eliminate RNases with dry heat (180 °C for 2 h).
12. PBST-DEPC: PBS-DPEC, 0.5% (v/v) Triton X-100.
13. 20x Saline-sodium citrate buffer (SSC): 0.3 M trisodium citrate, 3 M NaCl. Adjust pH to 7.0 with HCl. Treat with 0.1% DEPC overnight with shaking, and autoclave for 40 min. When required, dilute 20× SSC to the appropriate concentration as indicated using DEPC-treated water.
14. Peroxidase inactivation solution: 0.03% H_2O_2 in 4× SSC. Mix immediately before use.
15. Hybridization solution: 5× SSC, 50% formamide (molecular biology grade, deionized), 10% dextran sulfate, 1 mg/mL *Torula* RNA (Merck R6625; *see* **Note 3**), 1× Denhardt's Solution, (Merck D2532). Store at −20 °C.
16. Humid chamber.
17. Hybridization oven.

18. Hybridization Washing Buffer: 5× SSC, 50% formamide, 0.1% Tween-20.
19. Parafilm.
20. MAB-T buffer: 100 mM maleic acid, 150 mM NaCl. Adjust pH to 7.5 with NaOH, autoclave, and add 0.1% Tween-20.
21. Antibody blocking solution: Add 0.5% (w/v) Blocking reagent (Merck 11096176001) to MAB-T, and dissolve by heating at 60 °C with periodic shaking. Add 1% (w/v) bovine serum albumin (BSA).
22. Anti-digoxigenin antibody, peroxidase conjugated (Anti-DIG-POD, Merck 11207733910).
23. Tyramide-FITC can be purchased from commercial providers or prepared as follows [3]. Prepare Fluorescein-NHS ester solution at 10 mg/mL in dimethylformamide (DMFA) in a 15 mL Falcon tube. Because we buy a 100 mg package of NHS-Fluorescein, the easiest thing to do is to dissolve everything in 10 mL DMFA, and from there do two reactions, with 4 mL of the NHS-Fluorescein stock each (2 mL remain, discard them). Keep in the dark. Prepare a DMFA-triethylamine solution by addition of 50 μL of triethylamine to 5 mL of DMFA. Prepare a Tyramine solution by dissolving 50 mg of tyramine in 5 mL of DMFA-triethylamine solution. To obtain tyramide-FITC mix 4 mL fluorescein-NHS ester solution (in DMFA) and 1.37 mL tyramine solution (in DMFA-triethylamine) in a 15 mL tube, and incubate in the dark at room temperature for 2 h. Add 4.6 mL of 100% ethanol, and store in dark at −20 °C. Use as a 1:100 dilution as detailed in the WMISH working protocol.
24. Tyramide development solution: PBS pH 7.6, 0.1 M imidazole, adjust the pH after the addition of imidazole. Add 0.0015% H_2O_2 immediately before use.
25. 4′,6-diamidino-2-phenylindole (DAPI) solution: 1 μg/mL DAPI in PBS, from a 10 mg/mL DAPI stock solution prepared in dimethylformamide.
26. Mounting medium: 80% glycerol in PBS.
27. Coverslips.

2.4 Materials for WMISH of *F. hepatica* NEJs

1. DEPC-treated water.
2. PBS-DEPC.
3. PFA/PBS-DEPC.
4. Tabletop centrifuge.
5. Methanol (for molecular biology).

6. Holtfreter's solution: 59 mM NaCl, 6.7 mM KCl, 0.76 mM $CaCl_2$, 2.4 mM $NaHCO_3$. Treat with 0.1% DEPC overnight with shaking, and autoclave for 40 min.
7. Absolute ethanol, molecular biology grade.
8. 75% ethanol in Holtfreter's solution.
9. 50% ethanol in Holtfreter's solution.
10. Holtfreter's solution with 0.1% Triton X-100.
11. Proteinase K (20 mg/mL, molecular biology grade).
12. TEA buffer: 0.1 M Triethanolamine, pH 7. Dilute 1.3 mL of pure TEA in 100 mL of DEPC-treated water. Adjust the pH with HCl.
13. Acetic anhydride.
14. 20× SSC (*see* Subheading 2.3).
15. Hybridization buffer for WMISH: 5× SSC, 50% formamide (molecular biology grade, deionized), 1 mg/mL *Torula* RNA, 100 μg/mL Heparin, 1× Denhardt's solution (Merck D2532), 0.1% Tween 20, 0.1% CHAPS.
16. Thermomixer.
17. Hybridization washing buffer for WMISH: 2× SSC, 50% formamide, 0.1% Tween 20.
18. MAB-T (*see* Subheading 2.3).
19. Antibody blocking solution for WMISH: Add 1% (w/v) Blocking reagent (Merck 11096176001) to MAB-T, and dissolve by heating at 60 °C with periodic shaking. Then add 5% normal sheep serum and inactivate at 60 °C for 30 min.
20. Anti-digoxigenin antibody conjugated to alkaline phosphatase (Anti-DIG-AP, Merck 11093274910).
21. Alkaline phosphatase buffer (*see* Subheading 2.3).
22. Alkaline phosphatase buffer with 5 mM levamisole.
23. NBT/BCIP solution (*see* Subheading 2.3).
24. DAPI solution.
25. Mounting medium.
26. Microscope slides and cover slips.
27. Phenol–chloroform–isoamyl alcohol (25:24:1).
28. Chloroform.
29. HCl.

3 Methods

Carry out all procedures at room temperature, unless otherwise specified. Work under RNase free conditions. This includes cleaning the working bench and pipettes with RNase decontamination solution, followed by rinsing with 70% ethanol, using RNase free plastics, and using gloves at all stages. All solutions up to and including the hybridization step must also be RNase free.

3.1 In Vitro Transcription of DIG-Labeled Probes

Follow this protocol for each probe that will be synthesized. For each gene, you will require an antisense probe in order to detect its expression, as well as a control sense probe which is used in a parallel experiment and that should result in no signal. The antisense and sense probes are synthesized by transcribing the template DNA from downstream and upstream located promoters, respectively. Alternatively, an irrelevant probe of similar length (e.g., for a nucleic acid not present in *F. hepatica*) may be used as a control.

1. Set up the following reaction on ice and mix well by pipetting: Transcription Buffer 10×: 2 μL.

   DIG RNA labeling mix 10×: 2 μL.

   Appropriate viral RNA polymerase (e.g., T7, SP6, T3): 40 units.

   RNase inhibitor: 20 units.

   Template DNA containing the appropriate viral promoters: 1 μg.

   Water, nuclease free: enough for a 20 μL final reaction volume.
2. Incubate for 2 h at 37 °C in a water bath.
3. Add 2 units of DNase (RNase free) and incubate for 15 min at 37 °C to remove the template DNA.
4. Precipitate the probe overnight at −80 °C with 0.1 volumes of sodium acetate solution and 2.5 volumes of absolute ethanol.
5. Centrifuge at 12,000 × *g* for 20 min at 4 °C, wash the RNA pellet with 70% ethanol, dry the pellet and dissolve it in 30 μL of nuclease-free water.
6. Run 3 μL of each sample in a 1% agarose/TAE gel to confirm the integrity of the probe.

3.2 Dot-Blot Quantification of Probes

1. In a positively charged Nylon membrane, spot 1 μL drops of serial dilutions of the probes (undiluted, 1:10, 1:100, 1:1000) and of the control DIG-RNA probe, dilutions 10, 1.0, 0.1 and 0.01 ng/μL).

2. Dry for 1 h at 60 °C between two sheets of Whatman filter paper, and cross-link under UV light for 1 min in a transilluminator. Place the membrane in a plastic container.
3. Wash 5 min in Dot-Blot washing buffer with shaking.
4. Block 10 min in Dot-Blot blocking buffer with shaking.
5. Incubate with Anti-DIG-AP antibody 1:2000 dilution in Dot-Blot blocking buffer 20 min with shaking.
6. Wash two times, 10 min for each wash, with Dot-Blot washing Buffer.
7. Wash briefly in Dot-Blot detection buffer, with shaking.
8. Perform the detection with NBT/BCIP solution, without shaking (this is important to prevent extensive diffusion of the colored product). Usually the detection is ready after 5–10 min in the dark. Stop the reaction by washing with tap water.
9. Determine the approximate concentration of labeled probe by comparing the intensity of the signal in the dots of serial dilutions of your probes to the intensity of the signal in the dilutions of control DIG-labeled RNA of known concentrations.

3.3 Fluorescent ISH on Cryosections of F. hepatica Adults

1. Wash the adult specimens with PBS and fix them with at least 10 volumes of PFA/PBS at 4 °C overnight (*see* **Note 4**).
2. Wash the specimens three times, for 15 min each, in PBS-DEPC.
3. Incubate the specimens in 30% sucrose solution at 4 °C overnight.
4. Transfer the specimens to small (approximately 2 cm × 2 cm × 1 cm) containers constructed by folding aluminum foil and cover with OCT compound, to make the tissue blocks.
5. Flash-freeze the blocks by placing in a plastic container with approximately 1 cm of liquid nitrogen. Do not allow the liquid nitrogen get in direct contact with the OCT compound.
6. Keep the blocks at −20 °C until further use (they can be kept for at least 1 week in sealed plastic bags).
7. Remove the aluminum foil and cut into 10 μm thick cryosections using a cryostat. Pick up the sections using the RNase free coated glass slides. These slides can be kept until further use at −80 °C.
8. Thaw the slides for 30 min.
9. Refix the sections by covering the slides with PFA/PBS-DEPC for 15 min.

10. Transfer the slides to a Coplin jar and wash three times with PBST-DEPC (this will permeabilize the specimens) (*see* **Note 5**).
11. Inactivate endogenous peroxidase by incubating the slides for 30 min in peroxidase inactivation solution.
12. Wash the slides two times with 4× SSC (5 min for each wash).
13. Transfer the slides to a humid chamber and prehybridize by covering the slides for 10 min with approximately 500 μL of hybridization buffer at 55 °C in a hybridization oven.
14. Replace the hybridization buffer and incubate for 90 min at 55 °C in a humid chamber.
15. To prepare the DIG-labeled RNA probe, denature the probe at 80 °C for 3 min, spin centrifuge and place on ice. Add the probe to 500 μL of prewarmed hybridization buffer (55 °C) at a final concentration between 0.1 and 1 ng/μL. Add the hybridization buffer containing the probe to the sections. Cover the slides with a piece of Parafilm and incubate overnight at 55 °C.
16. Transfer the slides to a Coplin jar and remove the piece of Parafilm by filling the jar with 2× SSC, prewarmed at 55 °C (the pieces of Parafilm will float away after a few minutes).
17. Wash twice with prewarmed Hybridization Washing Buffer at 55 °C, 20 min per wash.
18. Wash twice with prewarmed 2× SSC at 55 °C, 10 min per wash.
19. Wash twice with prewarmed 0.2× SSC at 55 °C, 10 min per wash.
20. Transfer the slides to a humid chamber, and block by covering the slides with 500 μL of antibody blocking solution for 1 h.
21. Cover the slides with 500 μL of anti-DIG-POD (1:200) prepared in antibody blocking solution, and incubate for 2 h in the humid chamber.
22. Transfer the slides to a Coplin jar. Wash three times with MAB-T, 10 min each.
23. Transfer the slides to a humid chamber. Develop by covering the slides on a humid chamber with 500 μL of tyramide development solution containing a 1:100 dilution of FITC-Tyramide (prepare immediately before use), and allowing the reaction to proceed for 5–10 min.
24. Transfer the slides to a Coplin jar and wash three times with PBS for 10 min each.
25. Incubate with DAPI solution for 10 min.

26. Cover the slides with mounting medium and with a coverslip. The specimens can now be imaged by epifluorescence microscopy or confocal fluorescence microscopy.

3.4 WMISH of *F. hepatica* NEJ

Unless stated otherwise, include shaking during each incubation step.

1. Wash the NEJs with PBS and transfer to a 1.5 mL centrifuge tube. This will allow you to centrifuge the specimens after each step, so that they will settle quickly at the bottom of the tube (otherwise, many will be lost when pipetting out the solutions).
2. Fix the specimens with 1 mL of PFA/PBS-DEPC and leave overnight at 4 °C.
3. Replace with 1 mL of methanol and incubate for 10 min.
4. Exchange methanol and store at −20 °C. Specimens can now be stored for long periods (at least 6 months) at −20 °C or used directly for in situ hybridization.
5. Rehydrate the specimens by washing in an ethanol dilution series: 1 ml of 100% ethanol, 5 min; 1 mL of 75% ethanol/Holtfreter's solution; 10 min; 1 mL of 50% ethanol/Holtfreter's solution, 10 min.
6. Wash three times with 1 mL of Holtfreter's solution with 0.1% Triton X-100.
7. Incubate the specimens with Proteinase K at a final concentration 20 μg/mL in 1 mL of Holtfreter's solution at 37 °C for 5 min, without shaking.
8. Stop the treatment with 1 mL of ice-cold Holtfreter's solution.
9. Rinse twice in 1 mL of 0.1 M TEA buffer for 5 min (mix by swirling the tube from time to time).
10. Add 2.5 μL of acetic anhydride to the second wash and swirl from time to time. After 5 min add 2.5 μL more acetic anhydride, and incubate for 5 min.
11. Wash twice for 5 min with Holtfreter's solution.
12. Refix the specimens for 60 min with 1 mL of 4% PFA in Holtfreter's solution.
13. Wash two times for 20 min or longer in 1 mL of Holtfreter's solution, to remove the paraformaldehyde.
14. Prehybridize the specimens with 1 mL of hybridization buffer for WMISH at 60 °C for 10 min in a thermomixer at 600 rpm.
15. Replace the hybridization buffer for WMISH (prewarmed to 55 °C) and prehybridize for at least 2 h at 55 °C in a thermomixer at 600 rpm. Specimens can be left prehybridizing overnight.

16. To prepare the DIG-labeled RNA probe, denature the probe at 80 °C for 3 min, centrifuge to collect the sample at the bottom of the tube and place on ice. Add the probe to 500 μL of prewarmed Hybridization buffer (55 °C) at a final concentration between 0.1 and 1 ng/μL. Remove the buffer from the prehybridization step from the specimens and add the Hybridization buffer for WMISH containing the probe to the specimens. Hybridize at 55 °C overnight in a thermomixer at 600 rpm.
17. Carefully remove the probe and store at −80 °C. The probe can be reused at least three times.
18. Wash two times with 1 mL of prewarmed hybridization buffer for WMISH at 55 °C for 10 min in a thermomixer at 600 rpm. While changing solutions keep the tubes on the thermomixer to keep specimens at 55 °C.
19. Wash four times with 1 mL of prewarmed hybridization washing buffer for WMISH at 55 °C for 60 min in a thermomixer at 600 rpm.
20. Wash twice with MAB-T for 20 min (from now on, all steps are once again performed at room temperature).
21. Block for 1 h with 500 μL antibody blocking solution for WMISH.
22. Remove the solution and replace with 500 μL of antibody blocking solution for WMISH containing 1:2000 dilution of Anti-DIG-AP. Place on a rocker overnight at 4 °C.
23. Wash one time for 5 min with 1 mL of MAB-T.
24. Wash three times for 1 h with MAB-T.
25. Wash once for 5 min in 1 mL alkaline phosphatase buffer.
26. Wash once for 10 min in 1 mL alkaline phosphatase buffer with 5 mM levamisole (an inhibitor of endogenous alkaline phosphatase).
27. Perform the detection with 1 mL NBT/BCIP solution and follow the development of the reaction under a microscope (for highly expressed genes, the signal will be apparent after 15–30 min, whereas the reaction may proceed overnight for genes of low expression levels). When the chromogenic reaction is complete, wash the specimens with PBS and mount on a microscope slide with mounting medium and a coverslip.

4 Notes

1. We recommend using commercially available nuclease-free double distilled water for enzymatic reactions instead of preparing nuclease-free water in the laboratory by treatment with diethyl dicarbonate (DEPC), as traces of DEPC may inactivate enzymes.
2. These fragments can be obtained by cloning RT-PCR fragments into a suitable vector containing upstream and downstream viral promoters. We recommend cDNA fragment sizes between 0.5 and 1.5 kb. Although the signal will be lower, shorter probes may be used for highly expressed genes, and we have successfully performed ISH with probes as short as 0.2 kb. Once the sequence of the cloned fragments is confirmed by sequencing, the recombinant plasmids can be used as templates for PCR using primers with the sequence of the viral promoters, resulting in large amounts of DNA products that can be purified by standard methods to use as templates for in vitro transcription.
3. Commercially available *Torula* RNA has to be additionally purified. Dissolve at least 500 mg of *Torula* RNA at 10 mg/mL in DEPC-treated water (dissolve at 65 °C). Extract with one volume of phenol–chloroform–isoamyl alcohol (25:24:1), centrifuge 20 min at 10,000 × *g* at 4 °C, and transfer the aqueous phase to a new tube. Extract with one volume of chloroform and centrifuge 20 min at 10,000 × *g* at 4 °C, and transfer the aqueous phase to a new tube. Precipitate the RNA by adding 0.1 volume of 3 M sodium acetate pH 5.2 (RNase-free) and 2.5 volumes of 100% ethanol, and centrifuge 20 min at 10,000 × *g* at 4 °C. Wash the pellet twice with 70% ethanol (RNase-free). Remove all the ethanol with a pipette. Dry the pellet (usually around 5 min by placing at 37 °C), and dissolve in 1 mL of DEPC-treated water (heat for 5–10 min at 65 °C if needed). Dilute to 100–200 mg/mL and store in aliquots at −20 °C.
4. To prevent the contraction and curling of specimens, these can be placed in a tube with PBS on ice for 5 min, and then fixed by removing the PBS and adding boiling PFA/PBS-DEPC under a fume hood. Cool immediately on ice and continue with overnight fixation at 4 °C.
5. Additional permeabilization steps may be included, such as treating the slides for 10 min with 0.2 N HCl or with 1 μg/mL proteinase K, molecular biology grade.

References

1. Jin L, Lloyd RV (1997) In situ hybridization: methods and applications. J Clin Lab Anal 11:2–9
2. Cancela M, Acosta D, Rinaldi G, Silva E, Durán R, Roche L, Zaha A, Carmona C, Tort J (2008) A distinctive repertoire of cathepsins is expressed by juvenile invasive *Fasciola hepatica*. Biochimie 90:1461–1475
3. Hopman AH, Ramaekers FC, Speel EJ (1998) Rapid synthesis of biotin-, digoxigenin-, trinitrophenyl-, and fluorochrome-labeled tyramides and their application for in situ hybridization using CARD amplification. J Histochem Cytochem 46:771–777

Chapter 8

Evasion of Host Immunity During *Fasciola hepatica* Infection

Robin J. Flynn and Mayowa Musah-Eroje

Abstract

Fasciola hepatica, the common liver fluke, causes infection of livestock throughout temperate regions of the globe. This helminth parasite has an indirect lifecycle, relying on the presence of the mud snail to complete its transition from egg to definitive host (Beesley et al., Transbound Emerg Dis 65:199–216, 2017). Within the definitive host, the parasite excysts in the intestine forming a newly excysted juvenile (NEJ) and migrates via the peritoneal cavity to the liver. Disease resulting from infection can be acute or chronic depending on the host and the number of parasites present. Sheep may succumb to a fatal acute infection if the challenge of metacercariae is great enough. However, in cattle chronic disease is the most likely outcome with parasites surviving for long periods of time. Annual losses are estimated to be in the region of US$ 2000 million to the agricultural industry (Beesley et al., Transbound Emerg Dis 65:199–216, 2017). Management of the disease depends heavily on chemotherapy with triclabendazole being the drug of choice, consistent use for over 20 years has resulted in drug-resistant strains emerging worldwide (Beesley et al., Int J Parasitol 47:11–20, 2017). A more sustainable approach to control would be through vaccination and indeed a lead candidate has been identified, cathepsin L1. Despite these promising results the parasite continues to confound our own and host efforts to generate long-lasting and effective immunity. In this brief review we focus our attention on those mechanisms that the parasite utilises to circumvent the innate based defense mechanisms within the host.

Key words *Fasciola hepatica*, Immune evasion, Helminth, Immunomodulatory, Cathepsin, Innate immunity

1 Immunity to *F. hepatica*

F. hepatica immunity in ruminant hosts mirrors to large extent the response seen to Schistosome species. During experimental infection there is a brief phase of lymphocyte proliferation accompanied by IFN-γ production; thereafter a prolonged phase of IL-4 and initial antibody production follows. Coinciding with onset of patency there is a switch toward an anergic phenotype [3–5].

After emerging with the intestine invading NEJs must be sensed by the innate pattern recognition receptor (PRR) network. Evidence from murine models would suggest that the production

Martin Cancela and Gabriela Maggioli (eds.), *Fasciola hepatica: Methods and Protocols*, Methods in Molecular Biology, vol. 2137, https://doi.org/10.1007/978-1-0716-0475-5_8, © Springer Science+Business Media, LLC, part of Springer Nature 2020

of canonical type-2 cytokines IL-25, IL-33, and TSLP are essential at this juncture in initiating the first wave of innate immune responses. Eosinophilia is a core characteristic of the antihelminth response with multiple studies suggesting a sliding scale of importance in helminth clearance. In nematode infection eosinophilia is known to be nonessential in nematode infections for the expulsion of parasites [6]. Swartz et al. have shown that eosinophils play no role in *S. mansoni* infection parameters such as egg deposition, worm burdens, liver enzymes, and granuloma size or number [7]. In *F. hepatica* infection Bossaert et al. showed that eosinophil counts were significantly elevated in infected cattle within 4 weeks of infection and remained so during the course of a 16 week infection period [8]. Zhang et al. demonstrated the presence of biphasic eosinophilia in *F. hepatica* infected sheep, with the peaks occurring at weeks 4 and 9–10 postinfection [9]. The importance of eosinophilia was again demonstrated by Chauvin et al., who demonstrated a positive relationship between the total eosinophil count and the infective dose administered to sheep, signifying a correlation between immune response and intensity of infection [10]. Importantly their role in protective immunity is well supported; Doy et al. suggested a role for eosinophils in resistance developed in immune rats [11]. Immune rats facing a challenge infection showed an increase in eosinophils within the *lamina propria* of the small intestine. Van Milligen et al. described an ex vivo model of the rat gut during infection, in immune rats [12]. Again, eosinophil counts were elevated in the *lamina propria* of immune rats. When NEJs migrated into the mucosa of immune rats they were found to be coated with both IgG1 and IgG2a antibodies and eosinophils. Later work [13] showed that eosinophils were essential for protection in the same model. The presence of parasite-specific antibody would make ADCC the most likely method of killing NEJs. This work is supported by studies of various species placing ADCC at the center of protective immunity against *F. hepatica* NEJs in cattle [14, 15].

Macrophages elicited by helminth infection have been shown to diverge from the normal paradigm of classically activated—nitric oxide producing—antibacterial cells. Gordon summarized and outlined the mechanisms by which parasitic helminths can interact with MΦ, causing their alternative activation [16]. Alternatively activated MΦ (AAMΦ) are denoted by their production of polyamines, proline, and IL-10. The differential regulation of L-arginine by MΦ has allowed workers to distinguish between these two populations of cells. AAMΦ metabolize L-arginine (Arg-1) using the enzyme arginase. AAMΦ induced by parasite infections have been shown to express a unique panel of markers: the mannose receptor along with a number of unique molecules such as intelectins, resistin-like molecules (RELM), chitinases, or chitinase-like proteins [17]. To date AAMΦ have been found in infections with a wide variety of parasites including *S. mansoni* [18], *Taenia crassiceps* [19], *F. hepatica* [20], *Litomosoides sigmodontis* and *Nippostrongylus brasiliensis* [21], *Brugia malayi* [22], and *H. polygyrus* [23]. Numerous studies have shown that AAMΦ regulate the type-2 immune response in various helminth infections and help to limit immunopathology. However, the protective role of AAMΦ was shown by using *H. polygyrus* [23]. Infection of mice revealed an accumulation of AAMΦ into the intestine and surrounding these worms. Moreover, drug abbreviation of infection giving rise to immunity magnified this sterilizing immune response and macrophage depletion demonstrated that AAMΦ were central to curative response. Importantly, administration of an arginase-1 inhibitor demonstrated a direct effect of AAMΦ on worm viability measured via cytochrome oxidase. A direct effect of AAMΦ on *F. hepatica* viability has yet to be shown but roles in directing or contributing to the Th2 response during infection is well established in multiple species [24–26].

2 Mechanisms of Immune Evasion

Given the depth of information that is known about innate effector mechanisms, there is a corresponding trend for our knowledge regarding specifics of immune evasion to arise from study of the interactions between *F. hepatica* and the innate leukocytes. From herein we will discuss and explore the nature of these interactions and where known their function effects. One of the first in vitro studies of Immunomodulation resulting from *F. hepatica* infection was recorded in 1985 [27]. They reported that the ability of lymphocytes, from infected sheep, to proliferate was reduced even when stimulated with the mitogen, ConA. Similar interactions between leukocytes and excretory–secretory (ES) products were observed by Jefferies et al. [28–30]. They studied the effect of ES products on both human and ovine neutrophils and found that ES products caused neutrophils to polarize, migrate and induced morphological changes going from spherical to elongated type cells. They also demonstrated an ability of ES to reduce the oxidative burst of sheep and human neutrophils in response to PMA in a dose dependent manner. This work was one of the first to suggest that the parasite is capable of modulating aspects of the immune system to evade damage or destruction. ES products are a complex of multiple secreted proteins, both actively and passively. Refining the molecules within ES and defining their mode of action has become paramount to understanding parasite evasion and including key molecules in future vaccination plans. Below we discuss two major classes of parasite modulators, enzymatic and nonenzymatic modulators, giving an overview of the major details we have gleaned from studies to date.

3 Enzymatic Modulators

3.1 Cathepsins

The cathepsin cysteine protease family, containing cathepsin L1 (CL1) are the most clearly defined molecules from *F. hepatica* with immunomodulation capabilities. Early after the initial identification of CL1, Carmona et al. [31] demonstrated that *F. hepatica* CL1 could prevent eosinophil mediated ADCC killing of NEJs. CL1 was capable of cleaving antibody at the Fc-Fab junction, thus preventing cell attachment. Prowse et al. [32] demonstrated again CL1 directly modulates the expression of CD4 on lymphocytes by cleaving the receptor enzymatically. This effect could be reversed in the presence of a specific cathepsin inhibitor. Thus, at a direct level CL1 modulates immune function through its enzyme activities. Brady et al. [33] had earlier described a model of coinfection where *F. hepatica* would suppress mechanisms of defense that were specifically directed at *Bordetella pertussis.* This resulted in a loss of bacterial specific IFN-γ production and a delay in clearance of bacteria from the lungs. In follow up work, O'Neill et al. [34] demonstrated that injection of CL1 would have the same negative effect on *B. pertussis* immune responses as a *F. hepatica* infection. By use of knockout mice, they were able to show that this suppression was partially mediated by IL-4. In IL-4$^{-/-}$ mice IFN-γ levels were elevated in comparison to wild-type mice following injection of CL1, but still were significantly lower than in controls. Administration of a cathepsin enzyme inhibitor revealed that enzyme activity was required for the full suppressive effect. The enzymatic nature of *F. hepatica* CL1 was shown to suppress septic shock in vivo by Donnelly et al. [35]. Moreover, CL1 acted on TRIF and not surface bound TLR4 and use of both chemical inhibition and an active-site mutant CL1 confirmed reliance on protease activity. The requirement for active CL1 was against demonstrated in DCs [36], where CL1 caused partial maturation of DCs in vitro. A downstream functional effect was detectable in terms of attenuated Th17 responses when CL1-exposed DCs were used. Indicating there might be multiple routes to deviation from a Th1 or Th17 response that the parasite can use.

3.2 Peroxiredoxin

A second class of enzymes derived from ES products has also been well documented for their roles in host immunomodulation. Peroxiredoxin (formerly Thioredoxin Peroxidase) is a 2-cys redox enzyme which can traditionally protect DNA from redox damage [37]. It is weakly recognised by the host with antibodies against Prx declining into chronic infection [37]. This in itself may parallel the period of infection when Prx is most potent, at the point during which macrophage recruitment during NEJ invasion is highest. The effect of Prx on macrophages, resulting in AAMΦ, has been demonstrated in multiple species. In mice, Prx causes strong induction of arginase-1, FIZZ1 and Ym1 [20] while in ruminants it was shown that arginase-1 and IL-10 were upregulated by Prx [38]. In ruminants acidic mammalian chitinase (AMCase) was also identified as being upregulated following Prx exposure. While chitinases are ancient enzymes known to degrade chitin, commonly found in arthropods, there are no chitin-substrates in *F. hepatica*—which raises the question of its function. Importantly, Prx was shown to cause AAMΦ independent of IL-4/IL-13 which indicated a mechanism for the parasite to by-pass canonical type-2-signalling. Furthermore, when neutralized by immunization prior to infection it was revealed that Ym1, indicating AAMΦ in the peritoneal cavity, was reduced as was the subsequent IL-4 response [39], ultimately indicating a role for Prx-induced AAMΦs propagating a type-2 immune environment. While the enzymatic function of Prx is essential for its function, the precise mechanism by which it establishes the AAMΦ phenotype remains unknown and may yet present a viable route to *F. hepatica* control.

4 Nonenzymatic Modulators

Recently a number of parasite modulators have emerged that do not rely on enzymatic activity to polarize or subvert host immune effector mechanisms. However a common feature among these immunomodulators is their homology to host proteins with immune functions.

4.1 HDM

F. hepatica helminth defense molecule (FhHDM) was initially identified through a proteomic screen and phylogenetic analysis confirmed that it shares structural similarities with human LL-37, an antimicrobial peptide [40]. Initial characterization suggested that FhHDM could bind to LPS and block septic shock in vivo. Further details on the mechanism of action of FhHDM revealed FhHDM bound to lipids in the membrane was internalized and subsequently blocked antigen presentation on the MHC-II complex [41]. During the internalization phase it was shown that lysosomal acidification was blocked and this resulted in decreased inflammasome activation and subsequent IL-1β secretion [42]. The consequence of blocking antigen presentation within infection might allow for evasion of adaptive responses during infection; however IL-1β has more recently been shown to suppress the protective responses against intestinal *Trichuris muris* [43]. Thus, it is possible that while inhibiting antigen presentation benefits *F. hepatica* survival a benefit of blocking IL-1β remains to be uncovered.

4.2 TLM

TLM, TGF-like molecule, was first described from a screen of the *F. hepatica* genome. It presented with restricted expression, being highly expressed within the NEJ stage and low levels of expression within the adults. Initial experiments demonstrated that TLM retained similar qualities to TGF signalling in other worms and promoted viability and motility in vitro. Sulaiman et al. [15] later demonstrated that effects of TLM were not parasite restricted. Solid-phase binding assays demonstrated that TLM could indeed bind host TGF receptor complexes and resulted in activation of host STAT signalling. Phenotyping of macrophages exposed to TLM demonstrated a deviation from the AAMΦ spectrum with a significant increase in markers associated with a regulatory response including PD-1 and CTLA4. Ultimately, preexposure to TLM resulted in a reduction in macrophage-mediated ADCC killing of the NEJ parasite. This presents a clear pathway from stage-specific secretion of a modulator through to a host tissue specific.

5 Summary

We present here a brief overview of some of the best characterized modulators, enzymatic and nonenzymatic, their modes of actions and phenotypic effects. Recent evidence would suggest that our attention should shift to components of the tegumental coat. In recent studies the crude tegumental coat has been shown to inhibit mast cells [44] and DCs [45] in driving Th1 responses. Interestingly some of the effects of tegumental antigens have shown to be both mannose receptor dependent and independent [46, 47], indicating that the composition of the tegumental antigen is complex and will require much further study. Elucidating the mechanisms of action of *F. hepatica* evasion molecules will benefit vaccine development and future biotherapeutics.

Acknowledgments

Funding: BBSRC Awards BB/M018369/1, BB/L011530/1, BB/M018520/1 to RJF and a University of Nottingham Vice-Chancellor Scholarship to MME.

References

1. Beesley NJ, Caminade C, Charlier J, Flynn RJ, Hodgkinson JE, Martinez-Moreno A, Martinez-Valladares M, Perez J, Rinaldi L, Williams DJL (2017) Fasciola and fasciolosis in ruminants in Europe: identifying research needs. Transbound Emerg Dis 65:199–216
2. Beesley NJ, Williams DJL, Paterson S, Hodgkinson J (2017) Fasciola hepatica demonstrates high levels of genetic diversity, a lack of population structure and high gene flow: possible implications for drug resistance. Int J Parasitol 47(1):11–20

3. Flynn RJ, Mulcahy G (2008) The roles of IL-10 and TGF-beta in controlling IL-4 and IFN-gamma production during experimental Fasciola hepatica infection. Int J Parasitol 38 (14):1673–1680
4. Flynn RJ, Mulcahy G, Elsheikha HM (2010) Coordinating innate and adaptive immunity in Fasciola hepatica infection: implications for control. Vet Parasitol 169(3–4):235–240
5. Sachdev D, Gough KC, Flynn RJ (2017) The chronic stages of bovine Fasciola hepatica are dominated by CD4 T-cell exhaustion. Front Immunol 8:1002
6. Grencis RK (1997) Th2-mediated host protective immunity to intestinal nematode infections. Philos Trans R Soc Lond Ser B Biol Sci 352(1359):1377–1384
7. Swartz JM, Dyer KD, Cheever AW, Ramalingam T, Pesnicak L, Domachowske JB, Lee JJ, Lee NA, Foster PS, Wynn TA, Rosenberg HF (2006) Schistosoma mansoni infection in eosinophil lineage-ablated mice. Blood 108(7):2420–2427
8. Bossaert K, Jacquinet E, Saunders J, Farnir F, Losson B (2000) Cell-mediated immune response in calves to single-dose, trickle, and challenge infections with Fasciola hepatica. Vet Parasitol 88(1–2):17–34
9. Zhang WY, Moreau E, Hope JC, Howard CJ, Huang WY, Chauvin A (2005) Fasciola hepatica and Fasciola gigantica: comparison of cellular response to experimental infection in sheep. Exp Parasitol 111(3):154–159
10. Chauvin A, Moreau E, Boulard C (2001) Responses of Fasciola hepatica infected sheep to various infection levels. Vet Res 32(1):87–92
11. Doy TG, Hughes DL, Harness E (1978) Resistance of the rat to reinfection with Fasciola hepatica and the possible involvement of intestinal eosinophil leucocytes. Res Vet Sci 25 (1):41–44
12. Van Milligen FJ, Cornelissen JB, Hendriks IM, Gaasenbeek CP, Bokhout BA (1998) Protection of Fasciola hepatica in the gut mucosa of immune rats is associated with infiltrates of eosinophils, IgG1 and IgG2a antibodies around the parasites. Parasite Immunol 20 (6):285–292
13. Van Milligen FJ, Cornelissen JB, Bokhout BA (1999) Protection against Fasciola hepatica in the intestine is highly correlated with eosinophil and immunoglobulin G1 responses against newly excysted juveniles. Parasite Immunol 21 (5):243–251
14. Duffus WP, Franks D (1980) In vitro effect of immune serum and bovine granulocytes on juvenile Fasciola hepatica. Clin Exp Immunol 41(3):430–440
15. Sulaiman AA, Zolnierczyk K, Japa O, Owen JP, Maddison BC, Emes RD, Hodgkinson JE, Gough KC, Flynn RJ (2016) A Trematode parasite derived growth factor binds and exerts influences on host immune functions via host cytokine receptor complexes. PLoS Pathog 12 (11):e1005991
16. Gordon S, Taylor PR (2005) Monocyte and macrophage heterogeneity. Nat Rev Immunol 5(12):953–964
17. Nair MG, Guild KJ, Artis D (2006) Novel effector molecules in type 2 inflammation: lessons drawn from helminth infection and allergy. J Immunol 177(3):1393–1399
18. Herbert DR, Holscher C, Mohrs M, Arendse B, Schwegmann A, Radwanska M, Leeto M, Kirsch R, Hall P, Mossmann H, Claussen B, Forster I, Brombacher F (2004) Alternative macrophage activation is essential for survival during schistosomiasis and downmodulates T helper 1 responses and immunopathology. Immunity 20(5):623–635
19. Terrazas LI, Montero D, Terrazas CA, Reyes JL, Rodriguez-Sosa M (2005) Role of the programmed Death-1 pathway in the suppressive activity of alternatively activated macrophages in experimental cysticercosis. Int J Parasitol 35 (13):1349–1358
20. Donnelly S, O'Neill SM, Sekiya M, Mulcahy G, Dalton JP (2005) Thioredoxin peroxidase secreted by Fasciola hepatica induces the alternative activation of macrophages. Infect Immun 73(1):166–173
21. Nair MG, Gallagher IJ, Taylor MD, Loke P, Coulson PS, Wilson RA, Maizels RM, Allen JE (2005) Chitinase and fizz family members are a generalized feature of nematode infection with selective upregulation of Ym1 and Fizz1 by antigen-presenting cells. Infect Immun 73 (1):385–394
22. Nair MG, Cochrane DW, Allen JE (2003) Macrophages in chronic type 2 inflammation have a novel phenotype characterized by the abundant expression of Ym1 and Fizz1 that can be partly replicated in vitro. Immunol Lett 85(2):173–180
23. Anthony RM, Urban JF Jr, Alem F, Hamed HA, Rozo CT, Boucher JL, Van Rooijen N, Gause WC (2006) Memory T(H)2 cells induce alternatively activated macrophages to mediate protection against nematode parasites. Nat Med 12(8):955–960
24. Pesce JT, Ramalingam TR, Mentink-Kane MM, Wilson MS, El Kasmi KC, Smith AM, Thompson RW, Cheever AW, Murray PJ,

Wynn TA (2009) Arginase-1-expressing macrophages suppress Th2 cytokine-driven inflammation and fibrosis. PLoS Pathog 5(4): e1000371

25. Pesce JT, Ramalingam TR, Wilson MS, Mentink-Kane MM, Thompson RW, Cheever AW, Urban JF Jr, Wynn TA (2009) Retnla (relmalpha/fizz1) suppresses helminth-induced Th2-type immunity. PLoS Pathog 5 (4):e1000393
26. Ramalingam TR, Pesce JT, Mentink-Kane MM, Madala S, Cheever AW, Comeau MR, Ziegler SF, Wynn TA (2009) Regulation of helminth-induced Th2 responses by thymic stromal lymphopoietin. J Immunol 182 (10):6452–6459
27. Oldham G, Williams L (1985) Cell mediated immunity to liver fluke antigens during experimental Fasciola hepatica infection of cattle. Parasite Immunol 7(5):503–516
28. Jefferies JR, Barrett J, Turner RJ (1996) Immunomodulation of sheep and human lymphocytes by Fasciola hepatica excretory-secretory products. Int J Parasitol 26 (10):1119–1121
29. Jefferies JR, Corbett E, Barrett J, Turner RJ (1996) Polarization and chemokinesis of ovine and human neutrophils in response to Fasciola hepatica excretory-secretory products. Int J Parasitol 26(4):409–414
30. Jefferies JR, Turner RJ, Barrett J (1997) Effect of Fasciola hepatica excretory-secretory products on the metabolic burst of sheep and human neutrophils. Int J Parasitol 27 (9):1025–1029
31. Carmona C, Dowd AJ, Smith AM, Dalton JP (1993) Cathepsin L proteinase secreted by Fasciola hepatica in vitro prevents antibody-mediated eosinophil attachment to newly excysted juveniles. Mol Biochem Parasitol 62 (1):9–17
32. Prowse RK, Chaplin P, Robinson HC, Spithill TW (2002) Fasciola hepatica cathepsin L suppresses sheep lymphocyte proliferation in vitro and modulates surface CD4 expression on human and ovine T cells. Parasite Immunol 24(2):57–66
33. Brady MT, O'Neill SM, Dalton JP, Mills KH (1999) Fasciola hepatica suppresses a protective Th1 response against Bordetella pertussis. Infect Immun 67(10):5372–5378
34. O'Neill SM, Mills KH, Dalton JP (2001) Fasciola hepatica cathepsin L cysteine proteinase suppresses Bordetella pertussis-specific interferon-gamma production in vivo. Parasite Immunol 23(10):541–547
35. Donnelly S, O'Neill SM, Stack CM, Robinson MW, Turnbull L, Whitchurch C, Dalton JP (2010) Helminth cysteine proteases inhibit TRIF-dependent activation of macrophages via degradation of TLR3. J Biol Chem 285 (5):3383–3392
36. Dowling DJ, Hamilton CM, Donnelly S, La Course J, Brophy PM, Dalton J, O'Neill SM (2010) Major secretory antigens of the helminth Fasciola hepatica activate a suppressive dendritic cell phenotype that attenuates Th17 cells but fails to activate Th2 immune responses. Infect Immun 78(2):793–801
37. Sekiya M, Mulcahy G, Irwin JA, Stack CM, Donnelly SM, Xu W, Collins P, Dalton JP (2006) Biochemical characterisation of the recombinant peroxiredoxin (FhePrx) of the liver fluke, Fasciola hepatica. FEBS Lett 580 (21):5016–5022
38. Flynn RJ, Irwin JA, Olivier M, Sekiya M, Dalton JP, Mulcahy G (2007) Alternative activation of ruminant macrophages by Fasciola hepatica. Vet Immunol Immunopathol 120 (1–2):31–40
39. Donnelly S, Stack CM, O'Neill SM, Sayed AA, Williams DL, Dalton JP (2008) Helminth 2-Cys peroxiredoxin drives Th2 responses through a mechanism involving alternatively activated macrophages. FASEB J 22 (11):4022–4032
40. Robinson MW, Donnelly S, Hutchinson AT, To J, Taylor NL, Norton RS, Perugini MA, Dalton JP (2011) A family of helminth molecules that modulate innate cell responses via molecular mimicry of host antimicrobial peptides. PLoS Pathog 7(5):e1002042
41. Robinson MW, Alvarado R, To J, Hutchinson AT, Dowdell SN, Lund M, Turnbull L, Whitchurch CB, O'Brien BA, Dalton JP, Donnelly S (2012) A helminth cathelicidin-like protein suppresses antigen processing and presentation in macrophages via inhibition of lysosomal vATPase. FASEB J 26 (11):4614–4627
42. Alvarado R, To J, Lund ME, Pinar A, Mansell A, Robinson MW, O'Brien BA, Dalton JP, Donnelly S (2017) The immune modulatory peptide FhHDM-1 secreted by the helminth Fasciola hepatica prevents NLRP3 inflammasome activation by inhibiting endolysosomal acidification in macrophages. FASEB J 31(1):85–95
43. Alhallaf R, Agha Z, Miller CM, Robertson AAB, Sotillo J, Croese J, Cooper MA, Masters SL, Kupz A, Smith NC, Loukas A, Giacomin PR (2018) The NLRP3 Inflammasome suppresses protective immunity to gastrointestinal

Helminth infection. Cell Rep 23 (4):1085–1098

44. Vukman KV, Adams PN, Metz M, Maurer M, O'Neill SM (2013) Fasciola hepatica tegumental coat impairs mast cells' ability to drive Th1 immune responses. J Immunol 190 (6):2873–2879
45. Vukman KV, Adams PN, O'Neill SM (2013) Fasciola hepatica tegumental coat antigen suppresses MAPK signalling in dendritic cells and up-regulates the expression of SOCS3. Parasite Immunol 35(7–8):234–238
46. Aldridge A, O'Neill SM (2016) Fasciola hepatica tegumental antigens induce anergic-like T cells via dendritic cells in a mannose receptor-dependent manner. Eur J Immunol 46 (5):1180–1192
47. Ravida A, Aldridge AM, Driessen NN, Heus FA, Hokke CH, O'Neill SM (2016) Fasciola hepatica surface coat glycoproteins contain Mannosylated and phosphorylated N-glycans and exhibit immune modulatory properties independent of the mannose receptor. PLoS Negl Trop Dis 10(4):e0004601

Chapter 9

Immunomodulatory Effect of *Fasciola hepatica* Excretory–Secretory Products on Macrophages

Lorena Guasconi, Marianela C. Serradell, Diana T. Masih, and Laura S. Chiapello

Abstract

The liver fluke, *Fasciola hepatica*, infects a wide range of mammals including humans and leads to chronic disease. Like other helminths, *F. hepatica* migrates and survives in the host tissues after penetrating the intestinal wall to enter the peritoneal cavity, and then migrates through the liver before finally inhabiting the bile ducts. To avoid the antihelminthic immune response during migration, *F. hepatica* releases excretory–secretory products (FhESP) that exert various immunomodulatory effects, such as alternative macrophage activation or programmed cell death induction. Here, we describe the currently available techniques for studying macrophage activation and apoptotic cell death triggered by purified FhESP originating from the adult *F. hepatica* fluke.

Key words *Fasciola hepatica*, Immunosuppression, Macrophages, Excretory–secretory products, Alternative activation, Apoptosis

1 Introduction

Fasciolosis is a chronic infection caused by the helminth parasite *F. hepatica*, which inhabits the bile ducts for a long period of time, thereby demonstrating its ability to evade the host immune system [1]. The immunomodulatory properties of helminths are to a large extent attributed to excretory–secretory (ES) molecules in the form of proteins, peptides, glycans, lipids, and other small organic molecules that the parasites release when migrating and feeding inside their hosts [2–4]. Related to this, *F. hepatica* excretory–secretory products (FhESP) have shown wide immunomodulatory properties [5–14], which is crucial during the early stages of infection and the establishment of the chronic disease.

Several publications have demonstrated that FhESP promote alternatively activated macrophages (aaMΦ) with anti-inflammatory properties such as IL-10 and TGF-β secretion, and also L-arginine consumption by arginase activation [15–20]. Furthermore, FhESP

Martin Cancela and Gabriela Maggioli (eds.), *Fasciola hepatica: Methods and Protocols*, Methods in Molecular Biology, vol. 2137, https://doi.org/10.1007/978-1-0716-0475-5_9,

induce macrophage and eosinophil apoptosis, a programmed cell death that can also prevent the functioning of innate immune cells, thus allowing for parasite persistence and restricting the host immunopathology [21–23].

The protocols described below are useful tools for evaluating FhESP effects on macrophages by studying the cell activation and survival.

2 Materials

2.1 Preparation of *Fasciola hepatica* Excretory–Secretory Products (FhESP)

1. Phosphate buffered saline (PBS), pH 7.4, endotoxin-free.
2. High-flow YM10 membrane filters.
3. Bradford protein assay.
4. Columns containing detoxi-gel endotoxin removing gel.
5. Limulus amebocyte lysate test (Endosafe Times Charles River, Laboratories Wilmington).

2.2 Purification and Culture of Macrophages from Peritoneal Exudate Cells (PECs)

1. Six- to eight-week-old female BALB/c mice.
2. 70% ethanol.
3. Sterile forceps and small straight surgical scissors.
4. Styrofoam block and pins for euthanized mice mounting.
5. Sterile Pasteur pipettes (3 ml).
6. PBS.
7. Sterile 100 mM EDTA solution in distilled water.
8. Harvest medium: PBS, 0.1% fetal bovine serum (FBS), 5 mM EDTA.
9. Sterile collection tubes.
10. 0.4% Trypan blue solution in PBS.
11. RPMI 1640 medium.
12. Sterile 200 mM L-glutamine solution in distilled water.
13. 50 mg/ml gentamicin solution.
14. Complete RPMI medium: RPMI 1640 medium, 10% FBS, 2 mM glutamine, 50 μg/ml gentamycin.
15. Neubauer chamber cell counting.
16. Flat-bottomed 96-well or 48-well tissue culture plates.
17. FhESP.

2.3 Peritoneal Macrophage (pMΦ) Activation Assays

2.3.1 Measurement of Macrophage Cytokine Production in Culture Supernatants by Enzyme-Linked Immunosorbent Assay (ELISA)

1. Cytokine ELISA Kits.
2. Enhanced protein-binding ELISA plate.
3. Binding Solution: 0.1 M Na_2HPO_4, adjusted to pH 9.0 or to pH 6.0 depending on which cytokine is to be determined.
4. PBS/Tween: 0.5 ml of Tween 20 in 1 l PBS.
5. Blocking Buffer: 10% FBS in PBS.
6. Blocking buffer/Tween: 0.5 ml Tween 20 in 1 l blocking buffer.
7. Tetramethylbenzidine (TMB) substrate solution.
8. Microplate reader.

2.3.2 Arginase Activity Assay

1. 0.1% Triton X-100 in distilled water.
2. Protease inhibitor cocktail: 1.5 mM pepstatin, 0.08 mM aprotinin, 2 mM Leupeptin.
3. 10 mM $MnCl_2$ solution in distilled water.
4. 1 M Tris–HCl solution in distilled water, pH 7.5.
5. 0.5 M L-arginine solution in distilled water, pH 9.7.
6. Acid mixture: H_2SO_4, H_3PO_4, H_2O (1:3:7).
7. 9% α-isonitrosopropiophenone (ISPF) dissolved in 100% ethanol.

2.3.3 Cytometric Analysis of Macrophage MHC-II and Costimulatory Molecule Expression

1. Round-bottomed 96-well microtiter plates.
2. Purified anti-mouse CD16/32 antibody.
3. Fluorochrome-conjugated primary antibodies.
4. Flow cytometry staining buffer (FCSB): PBS pH 7.4, 3% FBS.
5. 1% paraformaldehyde in PBS.

2.3.4 Macrophage Antigen Presentation to CD4+T Lymphocytes

1. Peritoneal macrophages, pMΦ.
2. Mononuclear spleen cells (MSCs) from Balb/c mice.
3. Sterile wire mesh screens.
4. Red blood cell lysis buffer pH 7.3: 150 mM NH_4Cl, 10 mM $KHCO_3$, 0.1 mM Na_2EDTA.
5. 10 cm culture plates.
6. Magnetically activated cell sorting (MACS®) materials (Miltenyi Biotec): MACS® Columns (LS), MACS® Separators and anti-FITC MicroBeads.
7. Working Buffer: PBS pH 7.2, 0.5% bovine serum albumin (BSA), 2 mM EDTA (degassed) (*see* **Note 1**).
8. FITC-conjugated primary antibody.
9. 1% paraformaldehyde in PBS.

10. Flat-bottomed 96-well tissue culture plates.
11. Complete RPMI 1640 medium with 50 μM 2-mercaptoethanol.
12. Anti-mouse CD3 antibody.
13. Tritiated [^{3}H] thymidine.

2.4 Macrophage Apoptosis Assays

2.4.1 Cytometric Analysis of Hypodiploid DNA After Propidium Iodide (PI) Staining

1. PECs.
2. Round-bottomed 96-well microtiter plates.
3. 70% ethanol.
4. PBS, pH 7.4.
5. RNase Type I-A, 1 mg/ml in PBS.
6. PI stock solution: 0.5 mg/ml PI in water.
7. PI working solution (50 μg/ml): 1/10 PI stock solution in PBS (3 month durability).

2.4.2 Cytometric Analysis of Apoptotic Cells by Annexin V Staining

1. PECs.
2. Round-bottomed 96-well microtiter plates.
3. 5× annexin V binding buffer: 50 mM HEPES pH 7.4, 700 mM NaCl, 12.5 mM $CaCl_2$. Dilute the 5× annexin V binding buffer in distilled water to obtain the 1× annexin V binding buffer.
4. FITC-conjugated annexin V.
5. PI solution: 100 μg/ml PI in PBS.

2.4.3 Cytomorphologic Analysis of Macrophage Apoptosis by May–Grünwald–Giemsa Staining

1. PECs.
2. PBS.
3. Cytospin centrifuge.
4. Glass slides.
5. Filter cards.
6. Sample chambers.
7. Dyes: May–Grünwald's staining solution, composed of eosin and methylene blue solution in methanol. Giemsa's staining solution, composed of methylene blue, azure, and eosin.
8. Giemsa working solution: 5 ml of Giemsa dye in 50 ml of distilled water.

3 Methods

The FhESP are prepared following a procedure described by Diaz et al. [24], with some modifications:

3.1 Preparation of Fasciola hepatica Excretory–Secretory Products (FhESP)

1. Obtain live *F. hepatica* adult worms from the bile ducts of bovine livers, wash them with PBS and then incubate the worms in PBS pH 7.4 (1 worm/2 ml PBS) for 3 h at 37 °C (*see* **Note 2**).
2. Collect the supernatant and centrifuge for 30 min at 10,000 × *g* at 4 °C. Then, discard the pellet and concentrate the supernatant by using a high-flow YM 10 membrane filter.
3. Quantify protein concentration with Bradford assay.
4. To remove bacterial lipopolysaccharides (LPS) contamination, pass the FhESP preparation through a column containing detoxi-gel endotoxin removing gel. After endotoxin removal, the quantity of LPS present in FhESP is determined by using the Limulus amebocyte lysate test, whose level should be similar to that of the background and complete RPMI 1640 medium.
5. Store aliquots at −80 °C.

3.2 Purification and Culture of Macrophages from Peritoneal Exudate Cells (PECs)

Macrophage purification from PECs is performed according to procedures described by Zhang et al. [25] and Dittel et al. [26], with some variations:

1. Euthanize mice by rapid cervical dislocation (*see* **Notes 3** and **4**).
2. Soak each mouse abdomen with 70% ethanol and mount it on its back on a Styrofoam block.
3. Cut the outer skin of the peritoneum with scissors and forceps, and manually retract the abdominal skin to expose the intact peritoneal wall.
4. Make a small incision in the inner skin of the peritoneum with a sterile scissor, and dispense 5 ml of ice-cold harvest medium with a sterile plastic Pasteur pipette into the peritoneal cavity. Gently massage the peritoneum to dislodge any attached cells into the PBS solution, and while holding up the skin with a forceps, use the Pasteur pipette to collect the fluid from the cavity. Dispense the PECs into collection tubes maintained on ice (*see* **Note 5**).
5. Optional: the washing of the peritoneal cavity can now be repeated.
6. Centrifuge PECs for 10 min at 1500 × *g* at 4 °C. Discard the supernatant and resuspend the cell pellet in complete RPMI medium by gently tapping the bottom of the tube and pipetting up and down.
7. Determine cell viability by the trypan blue exclusion test (*see* **Note 6**).

8. To obtain macrophage monolayers, add PECs into 96-well (3×10^5 cells/well, in a final volume of 200 μl) or 48-well (5×10^5 cells/well, in a final volume of 500 μl) flat-bottomed tissue culture plates. Allow cells to adhere for 2–3 h at 37 ° C in an atmosphere of 5% CO_2.
9. Remove nonadherent cells by gently washing with warm PBS (at least three washes). More than 90% of the remaining adherent cells should be macrophages (*see* **Note** 7).
10. To investigate the FhESP effects on macrophage activation and survival, perform adherent macrophage cultures (in at least triplicate wells) in complete RPMI medium with the addition of FhESP at a concentration of 20 μg (macrophage activation assays) or 50 μg (apoptosis assays) protein/ml, or with endotoxin-free PBS (unstimulated controls) for 48 h.

    Optional: To investigate the role of innate cell receptors in FhESP-induced macrophage activation or apoptosis, preincubate pMΦ with chemical inhibitors or blocking antibodies of macrophage receptors for 30 min at 37 °C in 5% CO_2. Then, wash cells and incubate with FhESP 20 μg/ml or 50 μg/ml for 48 h [18, 19, 23].

3.3 Macrophage Activation Assays

3.3.1 Measurement of Macrophage Cytokine Production in Culture Supernatants by Enzyme-Linked Immunosorbent Assay (ELISA)

Supernatants from a 48 h culture of adherent macrophages with FhESP (20 μg/ml) or PBS (*see* **step 3** in Subheading 3.2), as well as from a 72 h culture of macrophage antigen presentation to CD4+T lymphocyte assays (*see* **step 3** in Subheading 3.3.4) are collected, and cytokine determination by ELISA assay is performed according to the manufacturer's protocol.

1. Dilute the purified anti-cytokine capture antibody to 1–4 μg/ ml in binding solution. Add 100 μl of diluted antibody to the wells of an enhanced protein-binding ELISA plate. Seal plate to prevent evaporation. Incubate overnight at 4 °C.
2. Wash ≥3 times with PBS/Tween.
3. Remove the capture antibody solution, and block nonspecific binding by adding 200 μl of Blocking buffer per well.
4. Seal plate and incubate at room temperature (RT) for 1–2 h.
5. Wash ≥3 times with PBS/Tween.
6. Add standards and samples (diluted in Blocking buffer/ Tween) at 100 μl per well.
7. Seal the plate and incubate it for 2–4 h at RT or overnight at 4 °C.
8. Wash ≥4 times with PBS/Tween.
9. Dilute the biotinylated anti-cytokine detection antibody to 0.5–2 μg/ml in Blocking Buffer/Tween. Add 100 μl of diluted antibody to each well.

10. Seal the plate and incubate it for 1 h at RT.
11. Wash ≥4 times with PBS/Tween.
12. Dilute the streptavidin-HRP conjugate to its pretitered optimal concentration in Blocking buffer/Tween. Add 100 μl per well.
13. Seal the plate and incubate it at RT for 30 min.
14. Wash ≥5 times with PBS/Tween.
15. Use TMB according to the instructions. Immediately dispense 100 μl into each well. Incubate at RT (5–80 min) for color development (*see* **Note 8**).
16. Read the optical density (OD) for each well with a microplate reader set to 405 nm.

3.3.2 Arginase Activity Assay

One of the main characteristics of alternatively activated macrophages is an alteration related to arginine in their metabolism. Classically activated macrophages express inducible nitric oxide synthase (iNOS), an enzyme that allows them to metabolize arginine into nitric oxide (NO) for microbial killing. In contrast, in alternatively activated macrophages, arginase activity is induced rather than iNOS, and arginine is converted to urea and ornithine, a precursor of polyamines and collagen [27]. Therefore, FhESP-induced alternatively activated macrophages may contribute to tissue repairing and immunosuppression [20] (*see* **Note 9**).

Arginase activity is measured as previously described by Corraliza et al. [28]:

1. Cultivate pMΦ in complete RPMI medium with FhESP (20 μg/ml) or PBS (unstimulated controls) (*see* **step 3** in Subheading 3.2). After 48 h, wash with PBS and lyse the cells by adding 50 μl 0.1% Triton X-100 containing 5 μg pepstatin, 5 μg aprotinin and 5 μg leupeptin as protease inhibitors.
2. Stir the mixture for 30 min at RT.
3. After lysing the cells, add 50 μl of 10 mM $MnCl_2$ and 50 mM Tris–HCl to activate the enzyme by heating for 10 min at 55 °C.
4. To initiate arginine hydrolysis, add 25 μl 0.5 M L-arginine pH 9.7, to a 25 μl aliquot of the previously activated lysate. Incubate this mixture at 37 °C for 45 min.
5. Stop the reaction with 400 μl of an acid.
6. Quantify urea production after the addition of 25 μl of 9% ISPF and heating at 100 °C for 45 min. After 10 min in the dark, measure the optical density (OD) at 540 nm in a microplate reader using 200 μl aliquots in flat-bottomed 96 well plate.
7. Construct a calibration curve (*see* **Note 10**). Arginase activity is expressed as mU, mU/10^6 cells or μg urea/μg protein.

3.3.3 Cytometric Analysis of Macrophage MHC-II and Costimulatory Molecule Expression

Peritoneal macrophages can be analyzed by flow cytometry as an F4/80 positive population in PEC suspensions. By combining fluorochrome-conjugated antibodies against MHC-II, CD80, CD86, CD40, PD-L1, PD-L2, etc., the macrophage activation profile and the ability to present antigens to T lymphocytes is evaluated.

1. To analyze FhESP effects on macrophage molecules surface expression, place PECs (3×10^5) on a 96-well round-bottomed microtiter plate and culture in complete RPMI medium plus FhESP (20 μg/ml) or PBS for 48 h, as described above.
2. *Optional*: To investigate the role of innate cell receptors in FhESP-induced macrophage molecule expression, preincubate PECs with chemical inhibitors or blocking antibodies of receptors for 30 min at 37 °C in 5% CO_2. Then, wash cells and culture them with FhESP 20 μg/ml or PBS for 48 h [18].
3. To block nonspecific Fc-mediated interactions, incubate cells with purified anti-mouse CD16/CD32 (0.5 μl/1×10^6 cells; final volume 100 μl in FCSB for 30 min at 4 °C.
4. Centrifuge cells at 400 × *g* for 5 min at 4 °C and discard the supernatant.
5. Conjugated primary antibody staining: Add 50 μl of the fluorescence-labelled primary antibodies diluted in FCSB. Combine the recommended quantity of each primary antibody in an appropriate volume of FCSB, so that the final staining volume is 100 μl/well (i.e., 50 μl of cell sample, 50 μl of antibody mix). Place samples in a vortex at low velocity.
6. Incubate for at least 30 min in the dark at 4 °C.
7. Wash the cells with 200 μl ice-cold FCSB and centrifuge at 400 × *g* for 5 min.
8. Discard the supernatant and fix the cells in 1% paraformaldehyde.
9. Before carrying out the analysis, centrifuge and resuspend cells in an appropriate volume of FCSB.
10. Analyze samples by flow cytometry (*see* **Note 11**).

3.3.4 Macrophage Antigen Presentation to CD4+T Lymphocytes

CD4+T lymphocytes can be purified using Magnetic-activated cell sorting (MACS®) according to the manufacturer's protocol.

1. Obtain a spleen cell suspension from BALB/c mice, by pressing spleens through wire-mesh screens to separate the cells. Remove the erythrocytes by incubating cells in red blood lysis buffer for 3–5 min and washing them with RPMI medium. Collect mononuclear spleen cells (MSCs) after a 6 h adherence culture of spleen cells in complete RPMI medium in 10 cm culture plates to remove adherent cells.

2. After preparation of single-cell suspension count cells, centrifuge the cell sample at 400 × *g* for 5 min at 4 °C (*see* **Note 12**).
3. Resuspend 10^7 total cells in 100 μl of working buffer and add 10 μl FITC-conjugate anti-CD4 antibody.
4. Mix well and incubate for 10 min in the dark at 4–8 °C (*see* **Note 13**).
5. Wash cells to remove unbound primary antibody by adding 1–2 ml of working buffer per 10^7 cells and centrifuge at 300 × *g* for 10 min.

   OptiConal: Repeat washing step.
6. Pipette off supernatant completely and resuspend cell pellet in 90 μl of working buffer per 10^7 total cells.
7. Add 10 μl of Anti-FITC MicroBeads per 10^7 total cells.
8. Mix well and incubate for 15 min at 4–8 °C (*see* **Note 13**).
9. Wash cells by adding 1–2 ml of working buffer per 10^7 cells and centrifuge at 300 × *g* for 10 min.
10. Pipette off supernatant completely and resuspend up to 10^8 cells in 500 μl of working buffer.
11. Place column in the magnetic field of a suitable MACS® Separator and prepare column by rinsing with 3 ml of working buffer.
12. Apply cell suspension onto the column.
13. Collect unlabeled cells which pass through and wash column with 3 ml of working buffer. Perform washing steps by adding working buffer three times (3 ml), each time waiting until the column reservoir is empty.
14. Collect total effluent. This is the unlabeled cell fraction.
15. Remove column from the separator and place it in a suitable collection tube.
16. Pipette 5 ml of working buffer onto the column. Immediately flush out fraction with the magnetically labeled cells by firmly applying the plunger supplied with the column.

    As result of this positive selection, more than 97% of pure T cells should be obtained, with a viability of 98%.
17. Culture pMΦ in flat-bottomed 96-well plates for 48 h with FhESP (20 μg/ml) or PBS, as described above.
18. Wash with RPMI and fix the macrophages with 1% paraformaldehyde, which are then used in subsequent cocultures.
19. Cocultures: Add purified CD4+T cells (3×10^5/well) to fixed macrophages and incubate 4 days with complete RPMI medium plus 50 μM 2-mercaptoethanol at 37 °C in 5% CO_2, in the presence of anti-CD3 antibody to activate polyclonal T cell proliferation.

20. Measure T cell proliferation by incorporation of [^{3}H] thymidine after the addition of 1 μCi of [^{3}H] thymidine to each well in the last 18 h of culture, by using a cell harvester and a liquid scintillation counter (*see* **Note 14**).
21. Measure cytokine production by T cells in supernatants of 72 h cultures by ELISA, as described above.

3.4 Macrophage Apoptosis Assays

Apoptosis is a carefully regulated suicide program that can be manipulated by pathogens, thereby representing one of the key factors in the survival strategy of the host [29]. FhESP-induced macrophage apoptosis can be analyzed by several methods, such as hypodiploid DNA quantification, annexin V assay or changes in the cellular and nuclear morphology.

3.4.1 Cytometric Analysis of Hypodiploid DNA After Propidium Iodide (PI) Staining

Apoptotic cells reveal a sub-G1 peak (G1: the level of cellular DNA at phase G1 of the cell cycle) on the profile of PI-stained cellular DNA as visualized by histograms in flow cytometry analysis. The percentage of hypodiploid DNA is evaluated following a procedure described by Nicoletti et al. [30], with some variations:

1. To analyze the FhESP effects on macrophage survival, place PECs (3×10^5) on 96-well round-bottomed microtiter plates, and culture in complete RPMI medium plus FhESP (50 μg/ml) or PBS for 24 or 48 h, as described above.
2. Collect the cells in a 1.5 ml tube and wash them with PBS to remove the culture medium (*see* **Note 15**).
3. Add 1 ml of ice-cold 70% ethanol fix and permeabilize cells. Incubate overnight at 4 °C.
4. Centrifuge the cells for 15 min at $3000 \times g$.
5. Wash with PBS (1 ml) and resuspend in 0.5 ml PBS. Then, add 0.5 ml of RNase to 0.5 cell sample, after which, gently mix 1 ml PI working solution (*see* **Note 16**).
6. Incubate cells in the dark at RT for 15 min and maintain them at 4 °C in the dark until being measured.
7. Measure the fluorescence of individual nuclei in an FACS flow cytometer, with the cell apoptotic nuclei being identifiable by their hypodiploid DNA content.

3.4.2 Cytometric Analysis of Apoptotic Cells by Annexin V Staining

Soon after initiating apoptosis, cells translocate the membrane phosphatidylserine (PS) from the inner face of the plasma membrane to the cell surface. Once on the cell surface, PS can be easily detected by staining with a fluorescent conjugate of Annexin V, a protein that has a high affinity for PS. Detection can be analyzed by flow cytometry. This approach can differentiate between apoptosis and necrosis by performing both Annexin V-FITC and PI staining according to a procedure described by Martin et al. [31], with some variations.

1. To analyze FhESP effects on macrophage survival, place PECs (3×10^5) on 96-well round-bottomed microtiter plates and culture in complete RPMI medium plus FhESP (50 μg/ml) or PBS for 24 or 48 h, as described above.
2. Collect the cells in a 1.5 ml tube and wash them with PBS to remove the culture medium (*see* **Note 15**).
3. Resuspend the cells in 100 μl of annexin V binding buffer (*see* **Note 17**).
4. Incubate the cells with FITC-conjugated annexin V (0.5 μl/ 2×10^5 cells) for 15 min at RT in the dark.
5. Add 400 μl of 1× annexin V binding buffer and 5 μl PI solution to the cell suspension (*see* **Note 18**).
6. Analyze cells using flow cytometry after incubation for 5–10 min (*see* **Note 19**).

   Early apoptotic cells are stained with Annexin V alone, whereas necrotic and late apoptotic cells are stained with both Annexin V and PI.

3.4.3 Cytomorphology Analysis of Macrophage Apoptosis by May–Grünwald–Giemsa Staining

Cytospins are often used to immobilize cells on glass microscope slides. Cytospin preparations stained with May–Grünwald–Giemsa reveal morphological changes typical of apoptosis, such as chromatin condensed against the nuclear periphery or the entire nucleus coagulated into small and dense balls. Cytospins are performed according to a procedure described by Kho [32], with some variations:

1. To analyze the FhESP effects on macrophage survival, place PECs (3×10^5) on 96-well round-bottomed microtiter plates, and culture in complete RPMI medium plus FhESP (50 μg/ ml) or PBS for 24 or 48 h, as described above.
2. Collect the cells in an eppendorf tube and wash them with PBS to remove the culture medium (*see* **Note 15**).
3. Resuspend the cells in PBS at a final concentration of 1×10^6 cells/ml (*see* **Note 20**).
4. Determine the number of viable cells by trypan blue exclusion.
5. Assemble the glass slides, filter cards, and sample chambers in the cytospin centrifuge, ensuring that the centrifuge is balanced.
6. Add 200 μl of cell suspension to each of the sample chambers and spin at $1000 \times g$ for 5 min (*see* **Note 21**).
7. Remove the slides, filter cards, and sample chambers carefully from the centrifuge. Discard filter cards and sample chambers, and allow to air to dry the slides prior to being stained.
8. Mark the position of the cells with a permanent marker on the underside of the slide.

9. Stain the cells with May-Grünwald working solution for 5 min.
10. Wash the cells with water and stain them with Giemsa working solution for 15 min (*see* **Note 22**).
11. Wash the cells with water, and dry the slides in the upright position at RT.
12. Analyze the morphological characteristics of the cells with an optical microscope.

4 Notes

1. To eliminate bubbles, pass the working buffer solution through a 0.22 μm filter.
2. Adult *F. hepatica* worms can be obtained from infected bovine livers from a local abattoir.
3. In all trials, we used 6- to 8-week-old female BALB/c mice, which were housed and cared for in the animal resource facilities of the institution where we work, following institutional guidelines. All experimental protocols were approved by the corresponding Animal Experimentation Ethics Committee.
4. Euthanasia should be performed by skilled technicians to avoid excessive bleeding into the peritoneal cavity. Furthermore, CO_2 euthanasia can also be used.
5. If visible blood contamination is detected, the contaminated sample should be discarded.
6. The determination of cell viability by trypan blue exclusion is carried out by adding 0.1 volumes of trypan blue solution to 0.9 volumes of cells in PBS, with unstained cells being viable.
7. The remaining adherent cells should be highly enriched for MΦ, and flow cytometry analysis should reveal greater than 90% of F4/80$^+$ cells.
8. Prepare a working concentration of TMB substrate solution within 15 min of use by mixing equal volumes of Substrate Reagent A and Substrate Reagent B.
9. We did not observe NO production in the culture supernatant of pMΦ stimulated with FhESP (20 μg/ml). Nitrite accumulation, an indicator of NO production, was measured using the Griess reagent [29].
10. A calibration curve must be constructed using increasing amounts of urea between 1.5 and 30 μg. In this case, 400 μl of the acid mixture and 25 μl ISPF are added to 100 μl of urea solution at the appropriate concentrations, and the procedure followed is as described above.

11. All staining should be done on ice or at 2–8 °C with minimal exposure to light. Appropriate negative controls should be used to verify specificity and to rule out background staining (using untreated cells). An irrelevant antibody of the same isotype and concentration should also be tested (control isotypes). In order to discriminate the specific population of pMΦ, the F4/80+ cells should be considered.
12. Work fast, keep the cells cold, and use precooled solutions. Volumes for magnetic labeling given below are for 10^7 total cells. When working with fewer than 10^7 cells, use the same volumes as indicated. When working with higher cell numbers, scale up accordingly all reagent volumes and total volumes (e.g., for 2×10^7 total cells, use twice the volume of all indicated reagent volumes and total volumes).
13. Working on ice requires an increased incubation time. Higher temperatures and/or longer incubation times lead to nonspecific cell labeling.
14. As an alternative to [^{3}H]-thymidine, dye dilution assays such as CFSE can be used for the measurement of cell proliferation by cytometric analysis.
15. Collect the cells and then, in order to recover cells that may have adhered, add cold PBS and use a scraper in each well. Collect the remaining cells and centrifuge for 10 min at $3000 \times g$ at 4 °C.
16. PI is a suspected carcinogen and should be handled with care.
17. Due to the calcium dependence of the Annexin V/PS interaction, it is critical to avoid buffers containing EDTA or other calcium chelators during Annexin V experiments.
18. A viability dye can be used to resolve these late-stage apoptotic and necrotic cells (Annexin V positive; viability dye positive) from the early-stage apoptotic cells (Annexin V positive; viability dye negative). To carry this out, PI (propidium Iodide) or 7-ADD (7-aminoactinomycin D) can be used.
19. Cells should be analyzed within 4 h of the initial incubation period, due to possible adverse effects on the viability of cells left in the presence of PI or 7-AAD for prolonged periods. Store cells at 2–8 °C and protect them from light until performing the analysis.
20. If the cell suspension is too diluted, there will not be enough cells immobilized on the slide after the cytospin centrifugation. If the cell suspension is too concentrated, the cells will tend to clump together and be poorly dispersed following the cytospin.
21. Parameter like cell density, loading volume and spin speed should be optimized for each cell line used.

22. The pH value is a very important factor in staining, so any change will lead to a wrong staining reaction. The range of the most suitable pH is 6.5 to 6.8.

Acknowledgments

This work was supported by SECyT-UNC No: 30720150100933CB, FONCyT PICT No 2015-1425 and CONICET, PIP No: GI 112201 501002 60. We would like to thank Dr. Paul Hobson, a native speaker, for revision of the manuscript.

References

1. Mehmood K, Zhang H, Sabir AJ, Abbas RZ, Ijaz M, Durrani AZ, Saleem MH, Ur Rehman M, Iqbal MK, Wang Y, Ahmad HI, Abbas T, Hussain R, Ghori MT, Ali S, Khan AU, Li J (2017) A review on epidemiology, global prevalence and economical losses of fasciolosis in ruminants. Microb Pathog 109:253–262
2. Ruyssers NE, De Winter BY, De Man JG, LoukasA, Herman AG, Pelckmans PA, Moreels TG (2008). Worms and the treatment of inflammatory bowel disease: are molecules the answer? Clin Dev Immunol 2008:567314. https://doi.org/10.1155/2008/567314
3. Navarro S, Ferreira I, Loukas A (2013) The hookworm pharmacopoeia for inflammatory diseases. Int J Parasitol 43:225–231
4. Shepherd C, Navarro S, Wangchuk P, Wilson D, Daly NL, Loukas A (2015) Identifying the immunomodulatory components of helminthes. Parasite Immunol 37(6):293–303
5. Cervi L, Rubinstein H, Masih DT (1996) Involvement of excretion–secretion products from *Fasciola hepatica* inducing suppression of the cellular immune responses. Vet Parasitol 61:97–111
6. Cervi L, Masih DT (1997) Inhibition of spleen cell proliferative response to mitogens by excretory–secretory antigens of *Fasciola hepatica*. Int J Parasitol 27:573–579
7. Cervi L, Rossi G, Cejas H, Masih DT (1998) *Fasciola hepatica*-induced immune suppression of spleen mononuclear cell proliferation: role of nitric oxide. Clin Immunol Immunopathol 87:145–154
8. Cervi L, Rossi G, Masih DT (1999) Potential role for excretory–secretory forms of glutathione-S-transferase (GST) in *Fasciola hepatica*. Parasitology 119:627–633
9. Cervi L, Cejas H, Masih DT (2001) Cytokines involved in the immunosuppressor period in experimental fasciolosis in rats. Int J Parasitol 31:1467–1473
10. Chapman CB, Mitchell GF (1982) Proteolytic cleavage of immunoglobulin by enzymes released by *Fasciola hepatica*. Vet Parasitol 11:165–178
11. Duffus WP, Franks D (1980) In vitro effect of immune serum and bovine granulocytes on juvenile *Fasciola hepatica*. Clin Exp Immunol 41:430–440
12. Heffernan M, Smith A, Curtain D, McDonnell S, Ryan J, Dalton JP (1991) Characterisation of a cathepsin-B proteinase released by *Fasciola hepatica* (liver fluke). Biochem Soc Trans 19:27S
13. Masih DT, Cervi L, Casado JM (1996) Modification of accessory activity of peritoneal cells from *Fasciola hepatica* infected rats. Vet Immunol Immunopathol 539:257–268
14. Flynn RJ, Irwin JA, Olivier M, Sekiya M, Dalton JP, Mulcahy G (2007) Alternative activation of ruminant macrophages by *Fasciola hepatica*. Vet Immunol Immunopathol 120:31–40
15. Flynn RJ, Mulcahy G (2008) Possible role for toll-like receptors in interaction of *Fasciola hepatica* excretory/secretory products with bovine macrophages. Infect Immun 76:678–684
16. Donnelly S, Stack CM, O'Neill SM, Sayed AA, Williams DL, Dalton JP (2008) Helminth 2-Cys peroxiredoxin drives Th2 responses through a mechanism involving alternatively activated macrophages. FASEB J 22:4022–4032
17. Walsh KP, Brady MT, Finlay CM, Boon L, Mills KHG (2009) Infection with a helminth parasite attenuates autoimmunity through TGF-b mediated suppression of Th17 and Th1 responses. J Immunol 183:1577–1586

18. Guasconi L, Serradell MC, Garro AP, Iacobelli L, Masih DT (2011) C-type lectins on macrophages participate in the immunomodulatory response to Fasciola hepatica products. Immunology 133:386–396
19. Guasconi L, Chiapello LS, Masih DT (2015) Fasciola hepatica excretory-secretory products induce CD4+T cell anergy via selective up-regulation of PD-L2 expression on macrophages in a Dectin-1 dependent way. Immunobiology 220(7):934–939
20. Motran CC, Silvane L, Chiapello LS, Theumer MG, Ambrosio LF, Volpini X, Celias DP, Cervi L (2018) Helminth infections: recognition and modulation of the immune response by innate immune cells. Front Immunol 9:664. https://doi.org/10.3389/fimmu.2018.00664
21. Serradell MC, Guasconi L, Cervi L, Chiapello LS, Masih DT (2007) Excretory-secretory products from *Fasciola hepatica* induce eosinophil apoptosis by a caspase-dependent mechanism. Vet Immunol Immunopathol 117:197–208
22. Serradell MC, Guasconi L, Masih DT (2009) Involvement of a mitochondrial pathway and key role of hydrogen peroxide during eosinophil apoptosis induced by excretory-secretory products from Fasciola hepatica. Mol Biochem Parasitol 163(2):95–106
23. Guasconi L, Serradell MC, Masih DT (2012) Fasciola hepatica products induce apoptosis of peritoneal macrophages. Vet Immunol Immunopathol 148(3-4):359–363
24. Diaz A, Espino AM, Marcet R, Otero O, Torres D, Finlay CM et al (1998) Partial characterization of the epitope on excretory–secretory products of *Fasciola hepatica* recognized by monoclonal antibody ES78. J Parasitol 84:55–61
25. Zhang X, Goncalves R, Mosser DM (2008) The isolation and characterization of murine macrophages. Curr Protoc Immunol. Chapter 14, Unit–14.1. https://doi.org/10.1002/0471142735.im1401s83
26. Ray A, Dittel BN (2010) Isolation of mouse peritoneal cavity cells. J Vis Exp 35:e1488. https://doi.org/10.3791/1488
27. Mosser DM, Zhang X (2008) Activation of murine macrophages. Curr Protoc Immunol 83(1):14.2.1–14.2.8
28. Corraliza IM, Campo ML, Soler G, Modollel M (1994) Determination of arginase activity in macrophages: a micromethod. J Immunol Methods 174:231–235
29. Chiapello LS, Baronetti JL, Garro AP, Spesso MF, Masih DT (2008) *Cryptococcus neoformans* glucuronoxylomannan induces macrophage apoptosis mediated by nitric oxide in a caspase-independent pathway. Int Immunol 20 (12):1527–1541
30. Nicoletti I, Migliorati M, Pagliacci M, Grignani F, Riccardi CA (1991) Rapid and simple method for measuring thymocyte apoptosis by propidium iodide staining and flow cytometry. J Immunol Methods 139:271
31. Martin SJ, Reutelingsperger CP, McGahon AJ et al (1995) Early redistribution of plasma membrane phosphatidylserine is a general feature of apoptosis regardless of the initiating stimulus: inhibition by overexpression of Bcl-2 and Abl. J Exp Med 182:1545
32. Koh CM (2013) Preparation of cells for microscopy using cytospin. Methods Enzymol 533:235–240

Chapter 10

Study of Eosinophil Apoptosis Induced by *Fasciola hepatica* Excretory–Secretory Products

Marianela C. Serradell, Lorena Guasconi, Laura Cervi, Laura S. Chiapello, and Diana T. Masih

Abstract

The excretory-secretory products released by the liver fluke *Fasciola hepatica* (FhESP) are in close contact with the immune system and have different immunomodulatory effects associated with the parasite virulence. The control of the early immune response is crucial for the establishment of the fluke in the host. Related to this, eosinophils (Eo) are implicated as effector cells in helminthic infections, and the induction of Eo apoptosis has been demonstrated to be a remarkable immunoevasion mechanism induced by the parasite. In this chapter, we describe different techniques to assay Eo apoptosis triggered by FhESP as well as the mechanisms involved in this phenomenon.

Key words *Fasciola hepatica*, Excretory-secretory products, Eosinophils, Apoptosis, Caspases, Oxidative stress, Immunoevasion

1 Introduction

Helminth infections are chronic diseases, which imply the development of strategies by the parasites in order to avoid the immune system [1, 2]. The helminth parasite *Fasciola hepatica* causes liver fluke disease or fasciolosis, affecting the health of humans, as well as sheep, cattle, and goats, among others. Fasciolosis causes losses in agriculture estimated at >US$2000 million per year, with its now increasing prevalence in humans [3]. During fasciolosis, juvenile parasites migrate through the host tissues and adults settle in the biliary ducts, being in contact with different immune cells [4]. One of the features of helminth infections is the development of eosinophilia [5]. A short time after the infection occurs, there is an increase in the number of eosinophils (Eo) in the blood and the migration of these cells to the site of infection [6]. This recruitment of Eo is partially due to the chemoattractant activity of parasite products released during infection [7]. Since helminths' size precludes phagocytosis, they must be eliminated by other mechanisms.

Martin Cancela and Gabriela Maggioli (eds.), *Fasciola hepatica: Methods and Protocols*, Methods in Molecular Biology, vol. 2137, https://doi.org/10.1007/978-1-0716-0475-5_10,

In these sense, Eo damage or kill larvae stages of helminth parasites mainly by using antibody-dependent cellular cytotoxicity mechanism (ADCC) and leaking toxic molecules from cytoplasmic granules [8].

Despite those interactions, the persistence of the parasite for many years provides evidence of its ability to prevent or downmodulate the inflammatory response in the infection site. During the parasite's migration, the rapid turnover of outer glycocalyx prevents an attack from granulocytes due to the antibodies being unable to recognize the surface of the parasite [9].

Furthermore, the larvae and adults release excretory–secretory products (ESP) which exert different immunomodulatory effects [10–13]. Moreover, FhESP induce macrophage and eosinophil apoptosis [14, 15], and the induction of this programmed cell death of effector cells has been described as a mechanism of immunosuppression during intracellular or extracellular parasite infections [16, 17].

Apoptosis may depend solely on caspase activation or, in addition, might also involve a mitochondrial phase during which the mitochondrial membrane potential ($\Delta\Psi$m) is lost with proapoptotic proteins, such as cytochrome c (cyt-c), being released. A diverse range of stimuli, including reactive oxygen species (ROS), induce mitochondrial depolarization thereby triggering cyt-c translocation and resulting in the activation of the caspase cascade [18].

In this chapter, we describe techniques to study rat Eo apoptosis triggered by purified FhESP from adult *F. hepatica* fluke. The protocols described are useful tools not only to studying the Eo apoptosis induced by these parasitic products, but also for the study of other proõapoptotic stimuli in different cellular type.

2 Materials

2.1 Fasciola hepatica Excretory-Secretory Product (FhESP) Preparation

See Chapter 9, Subheading 2.1.

2.2 Peritoneal Exudates Cells (PECs)

1. Ten to twelve-week-old male Wistar rats (*see* **Note 1**).
2. CO_2 euthanasia chamber (*see* **Note 2**).
3. 70% ethanol.
4. Sterile forceps and small, straight surgical scissors.
5. Styrofoam block and pins for euthanized mice mounting.
6. Sterile Pasteur pipettes (3 ml).
7. Phosphate-buffered saline (PBS).

8. Fetal bovine serum (FBS).
9. Ethylenediaminetetraacetic acid solution (EDTA).
10. Harvest medium: PBS, 0.1% FBS, 5 mM EDTA.
11. Sterile conical tubes (50 ml).
12. Hanks' balanced salt solution (HBSS).
13. Trypan blue solution: 0.4% trypan blue in PBS.
14. Neubauer chamber.
15. Optic microscopy.

2.3 Percoll® Gradient and Eo Purification

1. HBSS 10×.
2. HBSS 1×.
3. Percoll® colloidal silica particles (Density 1.130 g/ml).
4. Stock Isotonic Percoll® (SIP) (Density 1.123 g/ml) (*see* **Note 3** and Fig. 1a).
5. Percoll® with a density of 1.090 g/ml and Percoll® with density of 1.080 g/ml (*see* Fig. 1a).
6. PECs suspended in HBSS 1×.
7. Sterile conic tubes (15 ml).
8. Centrifuge.
9. Dye solutions: May–Grünwald and Giemsa.
10. Neubauer chamber and Trypan blue solution.
11. Optic microscopy.

2.4 Eo Culture

1. RPMI 1640 medium.
2. 200 mM L-glutamine solution.
3. 50 mg/ml Gentamicin solution.
4. Complete RPMI medium: RPMI 1640 medium supplemented with 10% FBS, 2 mM. glutamine, and 50 μg/ml gentamicin.
5. 96 well flat-bottomed tissue culture plates.
6. *F. hepatica* excretory-secretory products (FhESP).
7. Dexamethasone (positive apoptosis control) [19].
8. Inhibitors: rotenone (mitochondrial electron transport chain inhibitor [20]), genistein (tyrosine kinase inhibitor [21]), catalase (peroxide inhibitor [22]), superoxide dismutase-SOD (superoxide anion inhibitor [23]), diphenyleneiodonium chloride—DPI (NADPH oxidase inhibitor [20]), *N*-acetyl cysteine—NAC (antioxidant [24]), aminoguanidine (inhibition of the inducible nitric oxide synthase—iNOS [25]), Z-VAD-FMK (pan-caspase inhibitor [26]), staurosporine (protein kinase C inhibitor [27]).

a

$$\delta_{SIP} = \frac{V_{PP}\, x\, \delta_{PP} + V_M x\, \delta_M}{V_{PP} + V_M}$$

SIP: Stock Isotonic Percoll®

M (Medium): HBSS 10X δ HBSS 10X: 1,058 g/ml

PP: Pure Percoll® δ Pure Percoll®: 1,130 g/ml

$$\delta_{Percoll\ Solution} = \frac{V_{SIP}\, x\, \delta_{SIP} + V_M x\, \delta_M}{V_{SIP} + V_M}$$

SIP: Stock Isotonic Percoll® δ SIP : 1,123 g/ml

M (Medium): HBSS 1X δ HBSS 1X: 1,0046 g/ml

b

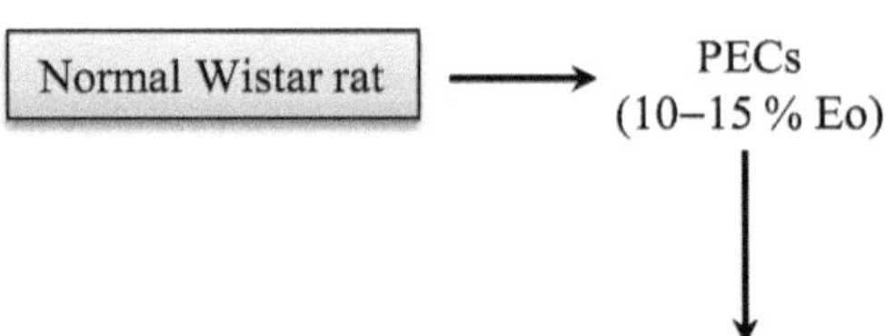

Discontinuos Percoll® gradient

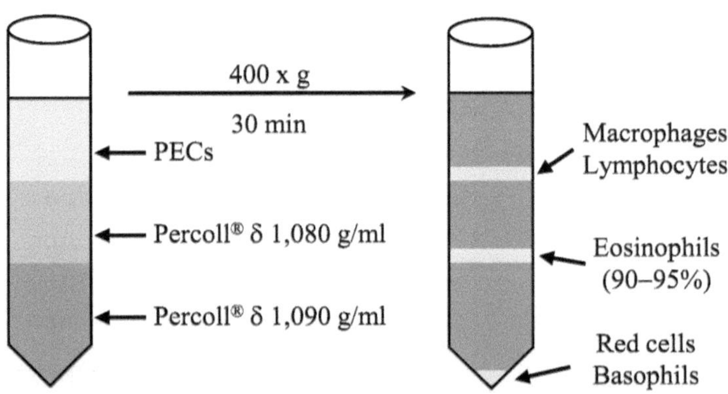

Fig. 1 Percoll® gradient and eosinophil purification. (**a**) Equations for preparation of the required solutions to performing the discontinuous Percoll® gradient. (**b**) Scheme showing the steps for Eo purification from rat peritoneal cells using a discontinuous Percoll® gradient

2.5 Apoptosis Assays

2.5.1 Morphologic Assessment

1. Purified cultured Eo.
2. PBS.
3. Cytospin centrifuge.
4. Glass slides.
5. Filter cards.

6. Sample chambers.
7. Dyes: May-Grünwald and Giemsa.
8. Optic microscopy.

2.5.2 Hypodiploid DNA Quantification by Propidium Iodide

1. Purified cultured Eo.
2. 96-well round bottom microplates.
3. 70% ethanol.
4. PBS, pH 7.4.
5. RNase Type I-A, 1 mg/ml in PBS.
6. Propidium iodide (PI) stock solution: 0.5 mg/ml PI in water.
7. PI working solution (50 μg/ml): 1/10 PI stock solution in PBS (3 month durability).
8. Flow cytometer.

2.5.3 Phosphatidylserine Surface Expression by Annexin V Assay

1. Purified cultured Eo.
2. 96-well round bottom microplates.
3. 5× annexin V binding buffer: 50 mM HEPES, 700 mM NaCl, 12.5 mM $CaCl_2$, pH 7.4.
4. FITC-conjugated annexin V.
5. 100 μg/ml PI solution.
6. Flow cytometer.

2.6 Study of Apoptotic Mechanisms

2.6.1 Measurement of Caspase Activity

1. Purified cultured Eo.
2. Commercial Caspase activity kit.
3. Protein quantification kit.
4. 96-well microplates.
5. Microplate reader with 405 nm filter.

2.6.2 Analysis of the Mitochondrial Membrane Potential ($\Delta\Psi m$)

1. Purified cultured Eo.
2. Tetramethylrhodamineethylester (TMRE) (*see* **Note 4**).
3. PBS.
4. Flow cytometer.

2.6.3 Quantification of Malondialdehyde Levels

1. Purified cultured Eo.
2. 15 M potassium chloride solution.
3. Sonicator.
4. 10% trichloroacetic acid (TCA) and 0.67% thiobarbituric acid (TBA).
5. Protein quantification kit.
6. Microplate reader with 535 nm filter.

2.6.4 Analysis of Reactive Oxygen Species (ROS) Production

1. Purified Eo.
2. 5-amino-2,3-dihydro-1,4-phthalazinedione (Luminol reagent).
3. 96-well black flat-bottom microplates.
4. Microplate plate luminometer.

2.6.5 Determination of Superoxide Production

1. Nitroblue tetrazolium 0.1% (NBT).
2. Phorbol 12-myristate 13-acetate (PMA).
3. HCl.
4. 1,4-dioxane.
5. Microplate reader with 560 nm filter.

2.6.6 Determination of Extracellular Hydrogen Peroxide Production

1. PRS buffer: 140 mM NaCl, 5.5 mM dextrose, 280 μM phenol red.
2. Peroxidase: 8.5 U/ml peroxidase (EC 1.11.1.7) in PBS, pH 7.0.
3. PMA.
4. NaOH 1 M.
5. PBS.
6. Microplate reader with 405 nm filter.

2.6.7 Determination of Intracellular Hydrogen Peroxide

1. Dye 2′, 7′-dichlorodihydrofluorescein diacetate (DCF).
2. PBS.
3. Flow cytometer.

2.6.8 Determination of Nitric Oxide (NO)

1. Sulfanilamide.
2. H_3PO_4.
3. (1-naphthyl)ethylenediamine.
4. $NaNO_2$.
5. H_2O.
6. Microplate reader with 540 nm filter.

3 Methods

3.1 Fasciola hepatica Excretory-Secretory Product (FhESP) Preparation

See Chapter 9, Subheading 3.1.

3.2 Peritoneal Exudates Cells (PECs)

1. Euthanize rats by CO_2 (*see* **Note 5**).
2. Soak each rat abdomen with 70% ethanol and mount it on a Styrofoam block on its back.

3. Cut the outer skin of the peritoneum with scissors and forceps. Manually retract the abdominal skin to expose the intact peritoneal wall.
4. Make an incision in the inner skin of the peritoneum with a sterile scissor, and dispense 15 ml of ice-cold harvest medium with a sterile plastic Pasteur pipette into the peritoneal cavity. Gently massage the peritoneum to dislodge any attached cells into the PBS solution, and while holding up the skin with a forceps, use Pasteur pipette to collect the fluid from the cavity. Dispense PECs in collection tubes kept on ice (*see* **Note 6**).
5. Repeat **step 4** twice.
6. Centrifuge PECs for 10 min at 400 × *g* at 4 °C. Discard supernatant and resuspend cell pellet in 2 ml of HBSS 1× (4–10 × 10^6 cells/ml) by gently tapping the bottom of the tube and pipetting up and down.
7. Determine cell viability by trypan blue exclusion test.

3.3 Percoll® Gradient and Eo Purification

Eo are purified from PECs of normal rats on a discontinuous Percoll® gradient [28] (*see* Fig. 1b).

1. To prepare the discontinuous Percoll® gradient, add 2 ml of Percoll® solution with a density of 1.090 g/ml to 2 ml with density of 1.080 g/ml, carefully overlaid on a 15 ml conic tube.
2. Add 2 ml of PEC suspension carefully over the Percoll® Gradient.
3. Centrifugate at 400 × *g* for 30 min at 4 °C (*see* **Note** 7).
4. Collect Eo from the interface between the Percoll® layers.
5. Resuspend cells in Complete RPMI medium (see below).
6. Count cells in a Neubauer chamber.
7. Determinate the percentage of purification by May-Grünwald-Giemsa staining (*see* **Note 8**).
8. Determinate the percentage of Eo viability using trypan blue exclusion (*see* **Note 9**).

3.4 Eo Culture

1. Resuspend fresh Eo at 2 × 10^6 cells/ml in complete RPMI.
2. Put aliquots of 100 μl of the cell suspension in 96-well flat bottom tissue culture plates.
3. Add 100 μl of the stimuli or medium alone (*see* **Note 10**).
4. Incubate at 37 °C and 5% CO_2 at different times depending on the test to be performed (*see* **Note 11**).
5. To evaluate the effect of apoptosis inhibitors, preincubate Eo for 30 min at 37 °C with DPI (50 μM), catalase (700; 1000; 1500 U/ml), SOD (900, 1800 U/ml), genistein (10^{-4} M), staurosporine (10^{-8} to 10^{-6} M), aminoguanidine (4 mM) or

Z-VAD-FMK (50 μM). In order to evaluate whether FhPES-induced apoptosis was mediated by oxidation, preincubate Eo for 1 h with the NAC antioxidant 5 mM.

3.5 Apoptosis Assays

We analyze the FhESP-induced Eo apoptosis by the following methods: changes in cellular and nuclear morphology; hypodiploid DNA quantification and annexin V assay.

3.5.1 Morphologic Assessment

Cytocentrifugated preparations are stained with May-Grünwald–Giemsa for optic microscopy analysis. The apoptotic cells are detected by morphological criteria. The normal rat Eo have ring-like nuclei, whereas apoptotic Eo are smaller than normal and have extensive condensed basophilic nuclei that are round in shape [29]. Cytospins are performed according to a procedure described by Kho [30], with some variations.

1. After 6 h of incubation at the different conditions, collect the cells (3×10^5) in an eppendorf tube and wash with PBS to remove the culture medium.
2. Resuspend the cells in PBS at a final concentration of 1×10^6 cells/ml.
3. Assemble the glass slides, filter cards, and sample chambers in the cytospin centrifuge, ensuring that the centrifuge is balanced.
4. Add 200 μl of cell suspension into each of the sample chambers and spin at $1000 \times g$ for 5 min.
5. Remove the slides, filter cards, and sample chambers carefully from the centrifuge. Discard filter cards and sample chambers, and allow to air-dry the slides prior being staining.
6. Mark the position of the cells with a permanent marker on the underside of the slide.
7. Stain the cells with May-Grünwald and Giemsa staining.
8. Analyze morphological characteristics of the cells with an optical microscope.

3.5.2 Hypodiploid DNA Quantification by Propidium Iodide

Apoptotic cells reveal a sub-G1 peak (G1: the level of cellular DNA at phase G1of the cell cycle) on the profile of PI-stained cellular DNA visualized by histograms in flow cytometry analysis. Therefore, the percentage of hypodiploid DNA is evaluated according to a procedure described by Nutku [29].

1. After 6 h of incubation at the different conditions, collect the cells (2×10^5) in an eppendorf tube and wash with PBS to remove the culture medium.
2. Add 1 ml of ice-cold 70% ethanol to fix and permeabilize cells. Incubate overnight at 4 °C.

3. Centrifuge the cells for 15 min at 1600 × *g*.
4. Wash with 1 ml of PBS and resuspend in 0.5 ml of PBS. Then, add 0.5 ml of RNAse to a 0.5 cell sample, followed after gentle mixing by 1 ml of PI solution (50 μg/ml) (*see* **Note 12**).
5. Incubate cells in the dark at room temperature for 15 min and kept at 4 °C in the dark until measured.
6. Measure the fluorescence of individual nuclei in a FACS flow cytometer. Eo are then identified on the basis of their forward and side scatter, and the cell debris are excluded. Eo apoptotic nuclei are distinguished by their hypodiploid DNA content.

3.5.3 Phosphatidylserine Surface Expression by Annexin V Assay

In early apoptosis, phosphatidylserine (PS) is relocated to the outer layer of the plasma membrane, with this appearing to be a general feature of apoptosis regardless of the initiating stimulus. FITC-labeled annexin V has been shown to provide a sensitive and rapid probe for apoptosis in a number of cell types, including B cells, neutrophils, cell lines, and vascular smooth muscle cells [31]. Detection can be analyzed by flow cytometry. This approach can differentiate apoptosis and necrosis when performing both annexin V-FITC and PI staining.

1. After 6 h of incubation at the different conditions, collect the cells in an eppendorf tube and wash with PBS to remove the culture medium.
2. Resuspend the cells in 100 μl of annexin V binding buffer (*see* **Note 13**).
3. Incubate the cells with FITC-conjugated annexin V (0.5 μl/ 2×10^5 cells) for 15 min at room temperature in the dark.
4. Add 400 μl of annexin V binding buffer and 5 μl PI solution (100 μg/ml in PBS) to cell suspension.
5. Incubate for 5–10 min and analyze cells using flow cytometry (*see* **Note 14**). Eo are gated on the basis of their forward and side light scatter. Early apoptotic cells are stained with annexin V alone, whereas necrotic and late apoptotic cells are stained with both annexin V and PI.

3.6 Study of Apoptotic Mechanisms

Eo apoptosis can be induced by different stimuli, such as the activation of CD95 receptors, oxidant components, or cellular stress factors. Different protein kinases have been involved in intracellular mechanisms of Eo survival and death [32]. This programmed cell death could be induced by an independent or dependent caspase pathway. Caspase-3 is an effector caspase, which typically functions downstream of caspase-8 and caspase-9, and directly activates enzymes that are responsible for DNA fragmentation. Caspase-8 activation occurs either apically in association with death receptors, or via the stress-induced signaling pathway involving mitochondria (also known as the intrinsic apoptotic pathway) [26]. Moreover, it is generally accepted that caspase-9 activation is strongly related to the apoptosis induced by stress [33]. Moreover, nitric oxide (NO) and reactive oxygen species (ROS) have often been associated with stress stimuli and the induction of apoptosis [34].

Apoptosis may depend solely on caspase activation or, in addition might also involve a mitochondrial phase during which the mitochondrial membrane potential (ΔΨm) is lost with pro-apoptotic proteins, such as cytochrome c (cyt-c), being released. A diverse range of stimuli, including reactive oxygen species (ROS), induce mitochondrial depolarization thereby triggering cyt-c translocation and resulting in the activation of the caspase cascade [26].

3.6.1 Measurement of Caspase Activity

Caspase-3 activity is measured by an Apoptosis Detection Colorimetric Kit, which includes DEVDp-nitroanilide as substrate for caspase-3 (*see* **Note 15**).

1. After Eo incubation by 3 h at the different conditions, collect the cells in an eppendorf tube and wash with PBS to remove the culture medium.
2. Resuspend the cells in 50 μl of lysis buffer (included in the kit) and incubate on ice for 15 min and centrifuge at 8000 × *g* for 10 min.
3. Collect the supernatants and transferred to a 96-well microplate.
4. Measured the protein concentration, and standardize all the samples the same amount of proteins.
5. Add 50 μl of reaction buffer (containing 10 mM DDT, included in the kit) and 5 μl of substrate (4 mM, included in the kit) and incubate at 37 °C for 3 h.
6. Measure the formation of *p*-nitroanilide every hour at 405 nm using a microplate reader. OD results are taken after the subtraction of the background.

3.6.2 Analysis of the Mitochondrial Membrane Potential (ΔΨm)

The highly fluorescent cationic lipophilic dyes TMRE is used to determine ΔΨm in intact cells by flow cytometry, since their retention is dependent on mitochondrial transmembrane potential [35].

1. After Eo incubation for 3 h under the different conditions, collect the cells in an eppendorf tube and wash with PBS to remove the culture medium.
2. Resuspend Eo in 50 nM of TMRE (*see* **Note 16**) and load for 30 min at 37 °C.
3. Analyze the stained cells by flow cytometry (*see* **Note 17**). TMRE is excited by the 488 nm laser and should be detected in the appropriate filter channel (peak emission is 575 nm).

3.6.3 Quantification of Malondialdehyde Levels

Lipid peroxidation is assessed from the formation of malondialdehyde (MDA), which is related to the oxidative damage suffered by cell membranes [36].

1. After Eo incubation by 6 h at the different conditions, collect the cells in an eppendorf tube and wash with PBS to remove the culture medium.
2. Resuspend the cells in 1.0 ml of 0.15 M potassium chloride solution and adjust to 1×10^6 cells/ml.
3. Sonicate samples, add 1.0 ml of 10% TCA and 1.0 ml of 0.67% TBA in each sample and mix well.
4. Centrifuge the aliquots of each vial at $1600 \times g$ for 10 min.
5. Decant the supernatant from each tube into separate tubes and place in a bath of boiling water for 10 min and then cool at room temperature.
6. Determinate the protein content of each sample.
7. Measure the absorbance of each aliquot at 535 nm. Express the rate of lipid peroxidation as nm of MDA formed per mg of protein, using a molar extinction coefficient of $1.56 \times 10^5\ M^{-1}\ cm^{-1}$.

3.6.4 Analysis of Reactive Oxygen Species (ROS) Production

ROS may work as signaling molecules to mediate various biological responses, including cell migration, growth, gene expression and apoptosis [37]. Since the two major species produced by the NADPH oxidase complex and mitochondria electron transport are O^{2-} and H_2O_2 [38], we also study their participation by measure the luminol-derived chemiluminescence, which have been reported to be elicited in response to the generation of an array of intracellular and extra cellular ROS.

1. Incubate the purified Eo (1×10^6) by 1 h with the different stimuli in 96-well black flat-bottom microplates at a final volume of 100 μl.
2. Add an equal volume of luminol (200 μM in PBS 1×).
3. Measure the chemiluminescence 30 min after incubation with the luminol. The data represent relative luminescence units (RLU) in an integration area of 10 s (area under the curve).

3.6.5 Determination of Superoxide Production

Superoxide anion (O^{2-}) is quantitatively determined by NBT reduction [39].

1. Incubate Eo for 1 h with the different stimuli.
2. Add NBT 0.1% with PMA at an optimal final concentration of 100 ng/ml and incubate in the dark for 30 min at 37 ° C, with 5% CO_2.
3. Stop the reaction with 0.4 ml of 0.1 N HCl.

4. Centrifuge the cells and extract the insoluble formazan twice with 1 ml of 1,4-dioxane.
5. Determinate the OD in supernatants at 560 nm in a microplate reader.

3.6.6 Determination of Extracellular Hydrogen Peroxide Production

The detection of H_2O_2 is performed by the phenol red oxidation microassay [39].

1. Incubate Eo (3×10^5) with the different stimuli in 96-well plates and left to stand at 37 °C in 5% CO_2 for 1 h.
2. Wash the cells once with PBS.
3. Replace the medium with 250 μl of PRS buffer and peroxidase in phosphate-buffered saline, pH 7.0 plus PMA (100 ng/ml), and incubate in the dark at 37 °C in 5% CO_2 for 45 min.
4. Stop the reaction with 10 μl of 1 M NaOH and read with a 595 nm filter. Express the results as nm of H_2O_2 released by 10^6 cells.

3.6.7 Determination of Intracellular Hydrogen Peroxide

The intracellular peroxide hydrogen quantification is made by a fluorescence probe oxidation and cytometric analysis. To measure intracellular H_2O_2, we use the cell-permeable redox-sensitive dye DCF because the nonfluorescent reduced form is converted into the fluorescent form when oxidized, thus allowing for detection by flow cytometry. DCF is oxidized by cellular H_2O_2, hydroxyl radicals, and other free radical products of H_2O_2, while it is relatively insensitive to oxidation by superoxide [40].

1. Incubate Eo for 1 h with the different stimuli.
2. Wash the cells with PBS, and then add 10 μM DCF for 20 min at 37 °C (*see* **Note 18**).
3. Analyze by flow cytometry for changes in fluorescence.

3.6.8 Determination of Nitric Oxide (NO)

The production of NO is determined through the measurement of nitrites in the supernatants of cell cultures using the Griess reaction [41].

1. Incubate Eo for 6 h with the different stimuli.
2. Add 200 μl of the Griess reagent (*see* **Note 19**) to 100 μl of the culture supernatant.
3. Incubate for 15 min at room temperature in the dark.
4. Determinate the reaction absorbance using a microplate reader at 540 nm.
5. Extrapolate the NO concentration of each sample based on the standard curve (*see* **Note 20**).

6. Production of NO^{2-} can be expressed as total concentration of NO^{2-} (i.e., in μM) or calculated as moles of NO^{2-} on the basis of cell number or amount of protein (i.e., in nmol/10^5 cells or nmol/mg).

4 Notes

1. Eight- or nine-week-old inbred female Wistar rats can also be used.
2. Compressed CO_2 gas in cylinders is the only recommended source of carbon dioxide as it allows the inflow of gas to the induction chamber to be controlled [42].
3. To make isotonic the 100% commercial Percoll® solution, to 9 parts of Percoll® add 1 part HBSS 10× (referred to as Stock Isotonic Percoll (SIP) in the chapter).
4. TMRE is a live cell stain and it is not compatible with fixation. TMRE is light sensitive and susceptible to photobleaching.
5. Without precharging the chamber, place the animals in the chamber and introduce 100% CO_2. A fill rate of about 10–30% of the chamber volume per minute with CO_2, added to the existing air in the chamber is appropriate to achieve a balanced gas mixture to fulfill the objective of rapid unconsciousness with minimal distress to the animals. Isoflurane overdose euthanasia could be used alternatively [42].
6. If visible blood contamination is detected the contaminated sample should be discarded.
7. Make sure the correct function of the centrifuge. The velocity is key in the success of purification.
8. Giemsa working solution: 5 ml of Giemsa dye in 50 ml of distilled water. The pH is a very important factor in staining, so any change will lead to wrong staining reaction. The limits of the most suitable pH are between 6.5 and 6.8. Stain the cells with May-Grünwald working solution for 5 min. Wash the cells with water and stain with Giemsa working solution for 15 min. Wash the cells with water, and dry the slides in upright position at room temperature.

   In general, it is expected to obtain a purity percentage close to 95%. In the peritoneum of normal rats an Eo percentage between 10% and 15% is usually found.
9. It is recommended to work with a viability percentage greater than 90%.
10. Stimuli for cell culture are 80 μg/ml of FhESP or dexamethasone (10^{-4} M, positive control of apoptosis).

11. In general, apoptosis assays are performed after 6 h of culture. In the case where the effect of rotenone is evaluated, the apoptosis is determined after 4 h because it had been demonstrated that this compound presents cytotoxicity effects after 6 h of incubation. The production of ROS, H_2O_2 and O^{2-}, is determined after 1 h incubation, and the ΔΨm is evaluated in cultures of 3 h.
12. PI is a suspected carcinogen and should be handled with care.
13. Due to the calcium dependence of the annexin–PS interaction, it is critical to avoid buffers containing EDTA or other calcium chelators during annexin V experiments.
14. Cells should be analyzed within 4 h after the initial incubation period due to adverse effects on the viability in the presence of PI or 7-AAD for prolonged periods. Store at 2–8 °C and protect from light until ready for analysis.
15. Caspase-3 is an effector caspase, which typically functions downstream of caspase-8 and caspase-9. However, there are a wide variety of commercial kits to determine the specific activity of other caspases.
16. Prepare a working TMRE solution by adding the appropriate volume of stock TMRE in appropriate culture media.
17. Removing media is not required for flow cytometry assays; however, an enhancement of TMRE fluorescence relative to background is achieved, when media is exchanged for PBS 0.2% FBS. The user should determine if washing away the media benefits their analysis.
18. This dye is not compatible with fixed samples. Stained cells must be measured live.
19. To prepare the Griess reagent mix equal volumes of 20 mg/ml sulfanilamide in 5% H_3PO_4 plus 2 mg/ml 0.1 N (1-naphthyl) ethylenediamine dihydrochloride in H_2O. For optimal results, add sulfanilamide before naphthylethylenediamine.
20. To perform a standard curve, set up serial dilutions of $NaNO_2$ in culture medium, covering a concentration range from 1 to 300 μM.

References

1. Scott P (2004) Immunoparasitology. Immunol Rev 201:5–8. https://doi.org/10.1111/j.0105-2896.2004.00194.x
2. Motran CC, Silvane L, Chiapello LS, Theumer MG, Ambrosio LF, Volpini X, Celias DP, Cervi L (2018) Helminth infections: recognition and modulation of the immune response by innate immune cells. Front Immunol 9:664. https://doi.org/10.3389/fimmu.2018.00664
3. Mehmood K, Zhang H, Sabir AJ, Abbas RZ, Ijaz M, Durrani AZ, Saleem MH, Ur Rehman M, Iqbal MK, Wang Y, Ahmad HI, Abbas T, Hussain R, Ghori MT, Ali S, Khan AU, Li J (2017) A review on epidemiology, global prevalence and economical losses of fasciolosis in ruminants. Microb Pathog 109:253–262. https://doi.org/10.1016/j.micpath.2017.06.006

4. Perez J, Ortega J, Bravo A, Diez-Banos P, Morrondo P, Moreno T, Martinez-Moreno A (2005) Phenotype of hepatic infiltrates and hepatic lymph nodes of lambs primarily and challenge infected with Fasciola hepatica, with and without triclabendazole treatment. Vet Res 36(1):1–12. https://doi.org/10.1051/vetres:2004047
5. Huang L, Appleton JA (2016) Eosinophils in Helminth infection: defenders and dupes. Trends Parasitol 32(10):798–807. https://doi.org/10.1016/j.pt.2016.05.004
6. Anthony RM, Rutitzky LI, Urban JF Jr, Stadecker MJ, Gause WC (2007) Protective immune mechanisms in helminth infection. Nat Rev Immunol 7(12):975–987. https://doi.org/10.1038/nri2199
7. Wildblood LA, Kerr K, Clark DA, Cameron A, Turner DG, Jones DG (2005) Production of eosinophil chemoattractant activity by ovine gastrointestinal nematodes. Vet Immunol Immunopathol 107(1–2):57–65. https://doi.org/10.1016/j.vetimm.2005.03.010
8. Makepeace BL, Martin C, Turner JD, Specht S (2012) Granulocytes in helminth infection—who is calling the shots? Curr Med Chem 19 (10):1567–1586
9. Hanna RE (1980) Fasciola hepatica: glycocalyx replacement in the juvenile as a possible mechanism for protection against host immunity. Exp Parasitol 50(1):103–114
10. Hillyer GV, Soler de Galanes M, Battisti G (1992) Fasciola hepatica: host responders and nonresponders to parasite glutathione S-transferase. Exp Parasitol 75(2):176–186
11. Cervi L, Masih DT (1997) Inhibition of spleen cell proliferative response to mitogens by excretory-secretory antigens of Fasciola hepatica. Int J Parasitol 27(5):573–579
12. Guasconi L, Serradell MC, Garro AP, Iacobelli L, Masih DT (2011) C-type lectins on macrophages participate in the immunomodulatory response to Fasciola hepatica products. Immunology 133(3):386–396. https://doi.org/10.1111/j.1365-2567.2011.03449.x
13. Berasain P, Carmona C, Frangione B, Dalton JP, Goni F (2000) Fasciola hepatica: parasite-secreted proteinases degrade all human IgG subclasses: determination of the specific cleavage sites and identification of the immunoglobulin fragments produced. Exp Parasitol 94 (2):99–110. https://doi.org/10.1006/expr.1999.4479
14. Guasconi L, Serradell MC, Masih DT (2012) Fasciola hepatica products induce apoptosis of peritoneal macrophages. Vet Immunol Immunopathol 148(3–4):359–363. https://doi.org/10.1016/j.vetimm.2012.06.022
15. Serradell MC, Guasconi L, Cervi L, Chiapello LS, Masih DT (2007) Excretory-secretory products from Fasciola hepatica induce eosinophil apoptosis by a caspase-dependent mechanism. Vet Immunol Immunopathol 117 (3–4):197–208. https://doi.org/10.1016/j.vetimm.2007.03.007
16. Luder CG, Gross U, Lopes MF (2001) Intracellular protozoan parasites and apoptosis: diverse strategies to modulate parasite-host interactions. Trends Parasitol 17(10):480–486
17. Zakeri A (2017) Helminth-induced apoptosis: a silent strategy for immunosuppression. Parasitology 144(13):1663–1676. https://doi.org/10.1017/S0031182017000841
18. Serradell MC, Guasconi L, Masih DT (2009) Involvement of a mitochondrial pathway and key role of hydrogen peroxide during eosinophil apoptosis induced by excretory-secretory products from Fasciola hepatica. Mol Biochem Parasitol 163(2):95–106. https://doi.org/10.1016/j.molbiopara.2008.10.005
19. Heasman SJ, Giles KM, Ward C, Rossi AG, Haslett C, Dransfield I (2003) Glucocorticoid-mediated regulation of granulocyte apoptosis and macrophage phagocytosis of apoptotic cells: implications for the resolution of inflammation. J Endocrinol 178 (1):29–36
20. Forkink M, Basit F, Teixeira J, Swarts HG, Koopman WJ, Willems PH (2015) Complex I and complex III inhibition specifically increase cytosolic hydrogen peroxide levels without inducing oxidative stress in HEK293 cells. Redox Biol 6:607–616. https://doi.org/10.1016/j.redox.2015.09.003
21. Wang Y, Sun HY, Liu YG, Song Z, She G, Xiao GS, Li GR, Deng XL (2017) Tyrphostin AG556 increases the activity of large conductance Ca(2+) -activated K(+) channels by inhibiting epidermal growth factor receptor tyrosine kinase. J Cell Mol Med 21 (9):1826–1834. https://doi.org/10.1111/jcmm.13103
22. Zhao Y, Carroll DW, You Y, Chaiswing L, Wen R, Batinic-Haberle I, Bondada S, Liang Y, St Clair DK (2017) A novel redox regulator, MnTnBuOE-2-PyP(5+), enhances normal hematopoietic stem/progenitor cell function. Redox Biol 12:129–138. https://doi.org/10.1016/j.redox.2017.02.005
23. De Cremer K, De Brucker K, Staes I, Peeters A, Van den Driessche F, Coenye T, Cammue BP, Thevissen K (2016) Stimulation of superoxide production increases fungicidal action of miconazole against Candida albicans biofilms. Sci Rep 6:27463. https://doi.org/10.1038/srep27463

24. Martinez-Losa M, Cortijo J, Juan G, Ramon M, Sanz MJ, Morcillo EJ (2007) Modulatory effects of N-acetyl-L-cysteine on human eosinophil apoptosis. Eur Respir J 30 (3):436–442. https://doi.org/10.1183/09031936.00073706
25. Chiapello LS, Baronetti JL, Garro AP, Spesso MF, Masih DT (2008) Cryptococcus neoformans glucuronoxylomannan induces macrophage apoptosis mediated by nitric oxide in a caspase-independent pathway. Int Immunol 20 (12):1527–1541. https://doi.org/10.1093/intimm/dxn112
26. Nutku E, Hudson SA, Bochner BS (2005) Mechanism of Siglec-8-induced human eosinophil apoptosis: role of caspases and mitochondrial injury. Biochem Biophys Res Commun 336(3):918–924. https://doi.org/10.1016/j.bbrc.2005.08.202
27. Peng J, Zheng H, Wang X, Cheng Z (2017) Upregulation of TLR4 via PKC activation contributes to impaired wound healing in high-glucose-treated kidney proximal tubular cells. PLoS One 12(5):e0178147. https://doi.org/10.1371/journal.pone.0178147
28. Sedgwick JB, Shikama Y, Nagata M, Brener K, Busse WW (1996) Effect of isolation protocol on eosinophil function: Percoll gradients versus immunomagnetic beads. J Immunol Methods 198(1):15–24
29. Nutku E, Aizawa H, Hudson SA, Bochner BS (2003) Ligation of Siglec-8: a selective mechanism for induction of human eosinophil apoptosis. Blood 101(12):5014–5020. https://doi.org/10.1182/blood-2002-10-3058
30. Koh CM (2013) Preparation of cells for microscopy using cytospin. Methods Enzymol 533:235–240. https://doi.org/10.1016/B978-0-12-420067-8.00016-7
31. Brumatti G, Sheridan C, Martin SJ (2008) Expression and purification of recombinant annexin V for the detection of membrane alterations on apoptotic cells. Methods 44 (3):235–240. https://doi.org/10.1016/j.ymeth.2007.11.010
32. Simon H, Alam R (1999) Regulation of eosinophil apoptosis: transduction of survival and death signals. Int Arch Allergy Immunol 118 (1):7–14. https://doi.org/10.1159/000024025
33. Kiraz Y, Adan A, Kartal Yandim M, Baran Y (2016) Major apoptotic mechanisms and genes involved in apoptosis. Tumour Biol 37 (7):8471–8486. https://doi.org/10.1007/s13277-016-5035-9
34. Chao TH, Chang MY, Su SJ, Su SH (2014) Inducible nitric oxide synthase mediates MG132 lethality in leukemic cells through mitochondrial depolarization. Free Radic Biol Med 74:175–187. https://doi.org/10.1016/j.freeradbiomed.2014.05.023
35. Jayaraman S (2005) Flow cytometric determination of mitochondrial membrane potential changes during apoptosis of T lymphocytic and pancreatic beta cell lines: comparison of tetramethylrhodamineethylester (TMRE), chloromethyl-X-rosamine (H2-CMX-Ros) and MitoTracker red 580 (MTR580). J Immunol Methods 306(1-2):68–79. https://doi.org/10.1016/j.jim.2005.07.024
36. Frijhoff J, Winyard PG, Zarkovic N, Davies SS, Stocker R, Cheng D, Knight AR, Taylor EL, Oettrich J, Ruskovska T, Gasparovic AC, Cuadrado A, Weber D, Poulsen HE, Grune T, Schmidt HH, Ghezzi P (2015) Clinical relevance of biomarkers of oxidative stress. Antioxid Redox Signal 23(14):1144–1170. https://doi.org/10.1089/ars.2015.6317
37. Ushio-Fukai M (2006) Localizing NADPH oxidase-derived ROS. Sci STKE 2006(349): re8. https://doi.org/10.1126/stke.3492006re8
38. Fialkow L, Wang Y, Downey GP (2007) Reactive oxygen and nitrogen species as signaling molecules regulating neutrophil function. Free Radic Biol Med 42(2):153–164. https://doi.org/10.1016/j.freeradbiomed.2006.09.030
39. Theumer MG, Lopez AG, Masih DT, Chulze SN, Rubinstein HR (2002) Immunobiological effects of fumonisin B1 in experimental subchronic mycotoxicoses in rats. Clin Diagn Lab Immunol 9(1):149–155
40. Forkink M, Willems PH, Koopman WJ, Grefte S (2015) Live-cell assessment of mitochondrial reactive oxygen species using dihydroethidine. Methods Mol Biol 1264:161–169. https://doi.org/10.1007/978-1-4939-2257-4_15
41. Archer S (1993) Measurement of nitric oxide in biological models. FASEB J 7(2):349–360
42. Resources UoIOoA (2017). https://animal.research.uiowa.edu/iacuc-guidelines-euthanasia. Accessed 02 July 2018

Chapter 11

Purification of Native *Fasciola hepatica* Fatty Acid-Binding Protein and Induction of Alternative Activation of Human Macrophages

Olgary Figueroa-Santiago and Ana Espino

Abstract

This chapter presents the different techniques to purify the native forms of *Fasciola hepatica* fatty acid-binding protein (Fh12) using size exclusion chromatography and isoelectric focusing (IEF). Also, it presents the procedure to study the immunological effect of the purified protein Fh12 using monocyte-derived macrophages (MDM) obtained from healthy human donors. For this purpose, I present the procedure to isolate and culture peripheral blood mononuclear cells (PBMCs) to generate alternatively activated macrophages (AAMΦ) by in vitro exposure to Fh12.

Key words *Fasciola hepatica*, Fatty acid-binding protein, Alternative activated macrophages, Monocyte-derived macrophages, Size exclusion chromatography, Ion exchange chromatography, Cell culture, Endotoxin removal

1 Introduction

The immunosuppressive effects of *Fasciola hepatica* have been associated with antigens secreted and excreted through the parasite tegument. One of these antigens is the proteins known as fatty acid-binding proteins (FABPs). FABPs are immunogenic proteins with molecular masses of 12–15 kDa. The FABP family members are expressed at different developmental stages of *F. hepatica* [1] and play essential roles in nutrients acquisition and parasite survival within the mammalian host.

Macrophages are innate immune cells that play an indispensable role in the primary response to pathogens but also they play a role in the resolution of inflammation and tissue homeostasis. Macrophages are integral members of the innate immune system

The original version of this chapter was revised. The correction to this chapter is available at https://doi.org/10.1007/978-1-0716-0475-5_18

Martin Cancela and Gabriela Maggioli (eds.), *Fasciola hepatica: Methods and Protocols*, Methods in Molecular Biology, vol. 2137, https://doi.org/10.1007/978-1-0716-0475-5_11, © Springer Science+Business Media, LLC, part of Springer Nature 2020, Corrected Publication 2020

and function as a first-line defense to detect, eliminate, or contain invading microbes and toxic macromolecules.

The activity of macrophages is controlled by specific signals that stimulate their development into discrete phenotypes, which differ regarding receptor expression, effector function, and cytokine secretions. The best-characterized phenotype is the classically activated macrophage (CAMΦ), which develops in response to proinflammatory cytokines such as interferon gamma (IFN-γ), microbial infection or bacterial products such as lipopolysaccharide (LPS) or CpG. In contrast, macrophages activated by an interleukin-4 (IL-4)/IL-13-dependent signal pathway are classified as alternatively activated macrophages (AAMΦ) [2]. It has been proposed that *F. hepatica* induces AAMΦ as a manipulative survival strategy which allows this parasite to survive in the mammalian host for long periods of time [2, 3]. A recent study demonstrated that Fh12 significantly increases arginase expression and activity, and induces the expression of the chitinase-3-like protein (CHI3L1) while at the same time downregulates the production of nitrite oxide and the expression of nitric oxide synthase 2 (NOS2) [4], indicating that Fh12 induces AAMØ. The study also demonstrated the downregulation of proinflammatory and inflammatory cytokines such as tumor necrosis factor-alpha (TNF-α), IL-12, and IL-1ßB. This chapter presents the protocols to purify the native form of Fh12 and the procedure to treat the cells with Fh12 to modulate human MDM and generate macrophages activated by an alternative phenotype.

2 Materials

2.1 Preparation of *Fasciola hepatica* Whole Worm Extract (FhWWE)

1. 10× phosphate buffered saline (PBS): Dissolve in 800 mL distilled H_2O, 25.6 g $Na_2HPO_4{\cdot}7H_2O$, 80 g NaCl, 2 g KCl, and 2 g KH_2PO_4. Adjust the pH to 7.4. Add dH_2O to the total volume of 1 L and autoclave.
2. 1× PBS containing the protease inhibitor PMSF (3 mM): Dilute 100 mL of the 10× PBS in a total volume of 1 L. Dissolve 0.174 g PMSF in 8 mL isopropanol. Add isopropanol to the total volume of 10 mL for a final concentration of 100 mM. Aliquot and store at −20 °C for several months. Dilute 0.3 mL of 100 mM PMSF in 9.7 mL 1× PBS for a final concentration of 3 mM. Store at 4 °C.
3. RPMI-1640: Dissolve 10.39 g RPMI-1640, 11.76 g D-glucose, and 25 mL 1 M HEPES. Add dH_2O to the total volume of 1 L and autoclave.
4. Wooden depressor.
5. 500 mL beakers.
6. Teflon homogenizer.

7. Ultracentrifuge.
8. BCA Protein assay kit.

2.2 Purification of Native 12 kDa Fatty Acid-Binding Protein

1. Sephadex G-50.
2. XK 26/100 column.
3. Funnel.
4. Pressure pump.
5. UV monitor.
6. Recorder.
7. 3% Sucrose: Dissolve 3 g in 80 mL dH_2O. Add dH_2O to the total volume of 100 mL.
8. Standard proteins: bovine serum albumin, ovalbumin, chymotrypsinogen, and ribonuclease.
9. 1× PBS containing 4 mM PMSF: Dilute 40 mL 100 mM PMSF in 960 mL 1× PBS for a final concentration of 4 mM. Store at 4 °C.

2.3 Antigen Separation Through Sephadex G50 Column

1. Test tubes.
2. 3% sucrose: Dissolve 3 g in 80 mL dH_2O. Add dH_2O to the total volume of 100 mL.
3. YM3 membrane.
4. High flow ultrafiltration system (AMICON).
5. 1% Glycine: Dissolve in 800 mL dH_2O. Add dH_2O to the total volume of 1 L.
6. Dialysis tubing cellulose membrane for retention of proteins with molecular weight of 12 kDa or higher.

2.4 Separation by Isoelectric Focusing (IEF)

1. 60 mL syringe.
2. Bio-Rad Rotofor.
3. Sealing Tape.
4. 1× PBS.
5. 1% Glycine.
6. Ampholytes: pH 3–10 and pH 5–7.
7. Acidic electrolytic solution: 0.1 M phosphoric acid (H_3PO_4). Dilute 3.57 mL of the 14 M H_3PO_4 in a total volume of 500 mL dH_2O.
8. Basic electrolytic solution: 0.1 M NaOH. Dissolve 4 g NaOH in 800 mL dH_2O. Add dH_2O to the total volume of 1 L.
9. Dialyzed Buffer: 10 mM Tris–HCl, pH 8, 15 mM NaCl. Dissolve 1.21 g Tris and 0.8766 g NaCl in 800 mL dH_2O and adjust the pH to 8. Add dH_2O to the total volume of 1 L.

2.5 Endotoxin Removal

1. Endotoxin removal kit.
2. Column stand.
3. Sodium azide reagent.
4. 1× PBS Sterile.
5. Micropipette.

2.6 Endotoxin Detection

1. Chambrex QCL-1000 Chromogenic LAL Endpoint Assay.
2. 25% acetic acid: Dilute 25 mL acetic acid in a total volume of 100 mL dH_2O.
3. Disposable endotoxin-free tubes.
4. 8-channel pipettor.
5. Disposable sterile microplates.
6. Reagent reservoirs.
7. Dry bath/multiblock heater at 37 °C ± 1 °C.
8. Vortex.
9. Spectrophotometer with a 405 nm filter.
10. Microplate reader.

2.7 Blood Sample Collection

1. Gloves and laboratory coat should be worn for venipuncture procedure.
2. Blood collection tubes: BD Vacutainer® CPT™—mononuclear cell preparation tubes with sodium citrate.
3. Needle package.
4. Centrifuge (swinging bucket) capable of generating at least 1500RCF (Relative Centrifugal Force).
5. Biological Safety Cabinet.
6. Sterile gauze.
7. Bandage.

2.8 Isolation of Human Peripheral Blood Mononuclear Cells

1. 1× PBS Sterile Ca^{2+}, Mg^{2+} free.
2. RPMI-1640 complete: media supplemented with 20 mM L-glutamine, 10% FBS heat-inactivated, and 1 mL of penicillin and streptomycin solution (50 U/mL)/100 mL of medium.
3. UV lamp.
4. 0.4% trypan blue.
5. Handheld counter.
6. Hemocytometer.
7. 50 mL sterile centrifuge tubes with caps.
8. Sterile serological pipettes (25, 10, 5, and 1 mL).
9. Centrifuge, swinging bucket.
10. Biological safety cabinet.

2.9 Human-Derived Macrophage and Fatty Acid-Binding Protein Treatment

1. 1× PBS Sterile, Ca^{2+}, Mg^{2+} free.
2. RPMI-1640 complete equilibrated at 37 °C.
3. 0.4% trypan blue.
4. Handheld counter.
5. Hemocytometer.
6. 50 mL sterile centrifuge tubes with caps.
7. Sterile serological pipettes (25, 10, 5, and 1 mL).
8. Sterile disposable 6-well culture plates.
9. Wet chamber.
10. Centrifuge with plaque adapter.
11. Scraper.
12. Biological Safety Cabinet.

2.10 Nitrite Oxide Assay

1. Griess Reagent System by Promega.
2. Sulfanilamide solution. Equilibrate at room temperature.
3. *N*-naphthylethylenediamine dihydrochloride (NED) solution. Equilibrate at room temperature.
4. Disposable sterile microplates (96 wells).
5. RPMI-1640 media.
6. 1.5 mL microtubes.
7. Multichannel pipette.
8. Sterile micropipette tips.
9. Microplate reader.
10. Spectrophotometer or filter photometer with 540 nm.

2.11 Arginase Activity

Bring the reagent to room temperature before assay. Use the reagent within 2 h after preparation.

1. Arginase Assay Kit.
2. Radioimmunoprecipitation assay (RIPA) Buffer: 150 mM NaCl, 1.0% *Octylphenoxy poly(ethyleneoxy)ethanol* (IGEPAL), 0.5% sodium deoxycholate, 0.1% SDS, 50 mM Tris, pH 8.0.
3. 5× Substrate Buffer: Combine 4 volume of Arginase Buffer and 1 volume of Mn solution. For each test, 10 μL of the 5× Substrate Buffer is needed.
4. 1 mM Urea Standard: Mix 24 μL of 50 mg urea/dl and 176 μL water.
5. Urea reagent: combine equal volume of Reagent A and Reagent B. For each test, 200 μL of the urea reagent is needed.
6. Clear bottom 96-well plates.
7. Pipetting devices.

8. Pipetting tips.
9. Microplate reader.
10. Spectrophotometer with 430 nm.

2.12 Cell Viability

1. Cell Proliferation Kit II (XTT) by Roche.
2. Thaw XTT labeling reagent and electron coupling reagent respectively in a water bath at 37 °C.
3. XTT labeling mixture: mix 50 μL of XTT labeling reagent and 1 μL electron coupling reagent. For each test, 100 μL XTT labeling mixture is needed.
4. Flat bottom 96-well plate.
5. Humidity chamber.
6. Microplate reader.
7. Spectrophotometer or filter photometer with 480 nm and 650 nm.

3 Methods

3.1 Preparation of FhWWE

1. Adult *F. hepatica* worms collected from the bile ducts of animals sacrificed at the slaughterhouse are sequentially washed with 1× PBS to eliminate all traces of blood and bile (*see* **Note 1**).
2. After washing, remove the solution and weight worms.
3. Resuspend washed worms in 1× PBS at a ratio of 1:2 (1 g per 2 mL of PBS).
4. Homogenize in a Teflon homogenizer using PBS supplemented with PMSF, during all process keep the worm suspension in ice. Repeat the process until you obtain a liquid mixture.
5. Ultracentrifuge the supernatants for 30 min at 30,000 × *g* at 4 °C to collect the supernatant and filter through 0.45 μm membrane.
6. Determine total protein concentration using the BCA protein assay kit.

3.2 Purification of Native 12 kDa Fatty Acid-Binding Protein

1. Resuspend the Sephadex G50 matrix in 1× PBS (1 g per 10 mL of PBS) by gradually adding the dry gel to the buffer and allow for swelling completely (*see* **Note 2**).
2. After the matrix is wholly swollen, allow the gel to settle, and degas the gel to remove air bubbles.
3. Fill the fourth part of the Sephadex G50 column (XK 26/100) with 1× PBS and gently pour 450 mL of degassed matrix suspension through the funnel into the column making sure a uniform package is obtained free of bubbles (*see* **Note 3**).

4. Close the packing reservoir and be sure that the tubbing is connecting to the pressure pump and from the pump to the column. Also, be sure that the column is connected to the UV monitor and the UV monitor to the recorder and the fraction collector. Set the UV monitor at 280 nm with range (AUFS) of 0.1 and the Recorder at cm/h.
5. Start to pack the column at a linear flow rate (LFr) of 5 cm/h. To convert LFr into volumetric flow rate (VFr), you should calculate the column area and the volumetric flow rate mL/h (*see* **Notes 4** and **5**).
6. Once the column has been packed, the packing reservoir will be removed, and the adapter placed and adjusted in close contact with the matrix. At this step wash, the column with 500 mL of 1× PBS, to be sure that the matrix has been packed to a 90 cm long bed.
7. Calibrate the column with standard proteins for molecular weight estimation. For this it is necessary to determine (a) void volume of the column (Vo), (b) final volume of the column (V_t), (c) distribution coefficient of each protein standard (K_{av}), and (d) calibration curve (log MW vs. K_{av}).
8. Determine the void volume (V_o) of the column, loading onto the column 3–4 mL of Blue Dextran (MW = 2,000,000 Da) at a concentration of 5 mg/mL and elute with 1× PBS.
9. Using a graduated cylinder collect and measure all volumes of PBS passing through the column and continue collecting volumes into a graduated cylinder until the first appearance of blue color (Blue dextran) is seen to elute from the column. The amount collected between the sample application and peak absorbance value is denoted as V_o (*see* **Note 6**).
10. Determine the final volume where π= 3.14 is the radius of the column using the following formula: $V_t = \pi r^2 L$, when $\pi = 3.14$, radius of the column (r^2) = $(1.3\ \text{cm})^2$, and length of the column (L) = (90 cm). The final volume (V_t) = $3.14 \times (1.3\ \text{cm})^2 \times 90\ \text{cm} = 477.6\ \text{cm}^3 = 477.6$ mL.
11. Determine the distribution coefficient of protein standards (K_{av}): bovine serum albumin (MW = 67,000 Da), ovalbumin (MW = 43,000 Da), chymotrypsinogen (MW = 25,000 Da), ribonuclease (MW = 13,700 Da). For K_{av} determination, dissolve 5–8 mg of each protein standard in 1 mL of PBS and load individually (4 mL) onto the column.
12. Set the pump to a maximal flow rate of 1 mL/min and start to elute the standard proteins. Samples should be loaded into the column through the three-way stopcock located in the tube connecting the buffer reservoir to the column using a syringe.

13. Add 1 mL of 3% sucrose to push all the proteins through the column and turn on the pressure pump. Monitor the purification using the UV monitor (absorbance at 280 nm, AUFS = 0.1).
14. Collect and measure all fluid eluted from the column from sample application up to reach the peak absorbance value (280 nm) of each protein. Plot DO 280 nm against "number of fractions."
15. Determine the calibration curve (*see* **Note** 7).

3.3 Antigen Separation Through Sephadex G50 Column

1. The protein concentration of the *Fh*WWE should not exceed 50 mg/mL.
2. The volume of *Fh*WWE to load onto the column in each run should range between 2% and 5% of V_t (*see* **Note 8**).
3. The flow rate for antigen separation is 1 mL/min (the same used for protein standard separation). The column separates the *Fh*WWE into two colored fractions. The first fraction is greenish color contain the higher MW proteins such as glutathione-S-transferase (*Fh*GST). The second fraction is of a reddish color, which contains the lower MW proteins (1.5×10^3 to 1.5×10^4) such as fatty acid-binding protein (*Fh*12).
4. To collect lower molecular weight fraction, select the tubes whose V_e are similar to the V_e of chymotrypsinogen and ribonuclease.
5. Analyze all these tubes by 15% SDS-PAGE.
6. Pool and concentrate the samples harvested up to an approximate volume of 25 mL using an ultrafiltration system YM-3 membrane.
7. Dialyze the concentrated fraction using dialysis membrane against 5 L of 1% glycine at 4 °C.

3.4 Protein Separation by Isoelectric Focusing (IEF)

1. Prior equipment assembly, equilibrate the anion and the cation exchange membrane in basic electrolytic solution and acidic electrolytic solution respectively for 24 h. After equilibration, keep both membranes from drying.
2. Place each ion exchange membrane inside the appropriate electrode between the two rubber circles. Immediately fill each electrode with the electrolyte solution. For anode or cathode exchange membrane (red bottom) add 30 mL acidic electrolytic solution and for cathode or anode exchange membrane (black bottom) add 30 mL basic electrolytic solution. Do not fill the chamber completely.

3. Connect the cooling water system to the cooling finger Rotofor. Place the electrodes and the focusing chamber on the cooling finger. Once the focusing chamber is assembly, proceed to IEF.
4. Equilibrate the system with dH_2O (twice 10 min, "power" 5 W) until the power supply reaches 2 mA: seal the collection ports (identified by two metal alignment pins) with seal tape.
5. Using a 60 mL syringe fill the focusing chamber (through the filling ports opposite to the collection port) with 55 mL of distilled water. Close the chamber. Change the chamber after each equilibration (use the IEF fraction collector).
6. Transfer 55 mL dialyzed fractions into a beaker and add the Biolyte ampholyte pH 3–10 at a final concentration 2%.
7. Fill the chamber with the 55 mL solution. Proceed to IEF: 12 W. Monitor IEF by writing down the volt and mA every hour (*see* **Note 9**).
8. Measure pH and the 280 nm absorbance of each tube and plot (absorbance 280 nm versus some fractions) and analyze fractions by 15% SDS-PAGE.
9. After SDS-PAGE select fractions containing only the *Fh*12 (MW = 12–15 kDa).
10. Pool recovered fractions and dialyze overnight at 4 °C against 5 L of dialyzed buffer to remove ampholytes.
11. Dialyze again using 5 L of 1% glycine overnight at 4 °C.
12. Repeat **steps 6–10** using Biolyte pH 5–7.
13. Collect samples containing only *Fh*12 between pH range 5.4 and 7.5 and pool the samples.
14. Dialyze against 1× PBS and concentrates on ultrafiltration using a YM-3 membrane.
15. Determine the protein concentration using the BCA method (*see* **Notes 10** and **11**).
16. Prepare aliquot in a small fraction of 500 μL and store at −20 °C until use.

3.5 Endotoxin Removal

1. Place the column upright in the column stand.
2. Remove the top cap first to prevent bubbles from being into the gel.
3. Allow storage solution to drain entirely from the column. Do not allow the column bed to dry.
4. Wash the column by adding 5 mL of cold regeneration buffer (Do not warm. Otherwise it will become cloudy) and let the buffer drain completely.

5. Set the flow rate at 0.25 mL/min or 10 drops per min by adjusting the flow speed control.
6. Repeat the wash step two more times to make the system endotoxin free (*see* **Note 12**).
7. Equilibrate the column by adding 6 mL of equilibration buffer and let the buffer drain entirely at a speed of 0.5 mL/min (*see* **Note 13**).
8. Repeat the equilibration step two more times.
9. Close the flow speed control after column equilibration.
10. Apply the sample to the column.
11. Set the flow rate at 0.25 mL/min or 10 drops per min by adjusting the flow speed control.
12. Collect sample after a void volume of 1.5 mL (*see* **Note 14**).
13. After the sample ultimately gets in the column, add 1.5 mL of sterile 1× PBS to recover the sample.
14. Pool the fractions containing protein sample and detect the endotoxin levels (*see* **Notes 15** and **16**).
15. For column storage, wash the column with 10 mL of equilibration buffer and allow the column to drain thoroughly.
16. Add 1.5 mL of regeneration buffer supplemented with 0.02% sodium azide and stored at 2–8 °C. Do not freeze.

3.6 Endotoxin Detection

1. Prepare a solution containing 1.0 EU/mL endotoxin by diluting 0.1 mL of the endotoxin stock solution with $(X - 1)/10$ mL of LAL reagent water in a suitable container, where X equals the endotoxin concentration of the vial. For example, If the concentration of the stock solution is 23 EU/mL, then dilute 0.1 mL of the endotoxin stock solution with 2.2 mL LAL Reagent Water ((23–1)/10).
2. Vortex vigorously for at least 1 min before proceeding.
3. Transfer 0.5 mL of 1.0 EU/mL into 0.5 mL of LAL Reagent Water in a suitable container and label 0.5 EU/mL. This solution should be vigorously vortexed for at least 1 min before use.
4. Transfer 0.5 mL of the 1.0 EU/mL into 1.5 mL of LAL Reagent Water in a suitable container and label 0.25 EU/mL. This solution should be vigorously vortexed for at least 1 min before use.
5. Transfer 0.1 mL of the 1 EU/mL into 0.9 mL of LAL Reagent Water in Water in a suitable container and label 0.1 EU/mL. This solution should be vigorously vortexed for at least 1 min before use.

6. After the standard preparation, preequilibrate the microplate at 37 °C ± 1 °C in the heating block adapter.
7. While leaving the microplate at 37 °C ± 1 °C, carefully dispense 50 μL of sample or standard into the appropriate microplate well (*see* **Note 17**).
8. At time $T = 0$, add 50 μL of LAL to the first microplate well, or the first column of microplate wells if using a multichannel pipettor and reagent reservoir. Begin timing as the LAL is added (*see* **Note 18**).
9. Once the LAL has been added to all wells containing sample or standard, briefly remove the microplate from the heating block adapter and repeatedly tap the side of the plate to facilitate mixing.
10. Return the plate on the heating block adapter and replace cover.
11. At $T = 10$ min, add 100 μL of substrate solution (prewarmed to 37 °C ± 1 °C) follow the recommendation in **Note 17**. Repeat **steps 9** and **10**.
12. At $T = 16$ min, following the recommendation in **Note 17**, add 50 μL of stop reagent (25% acetic acid).
13. Following the addition of the stop reagent, remove the plate and repeatedly tap it on the side.
14. Read the absorbance at 405 nm.

3.7 Blood Sample Collections

1. Select the tubes appropriated for samples desired. Tubes should be at room temperature and labeled adequately for patient identification.
2. Open needle package but do not remove needle shield. Thread the needle onto the holder.
3. Insert the tube into the holder. Leave in this position.
4. Select site for venipuncture.
5. Apply tourniquet. Prepare venipuncture site with an appropriate antiseptic. Do not palpate venipuncture area after cleansing. Allow the site to dry.
6. Remove the needle shield. Perform venipuncture with the patient's arm in a downward position and tube stopper uppermost; this reduces the risk of backflow of any anticoagulant into the patient's circulation.
7. Push the tube onto the needle, puncturing diaphragm of the stopper.
8. Remove tourniquet as soon as blood appears in the tube, within 2 min of venipuncture. Do not allow contents of the tube to contact the stopper or the end of the needle during the procedure (*see* **Note 19**).

9. When the first tube has filled to its stated volume, remove it from the holder.
10. Place succeeding tubes in the holder, puncturing the diaphragm to initiate flow.
11. Invert the tube 8 to 10 times to mix the anticoagulant additive with blood. Do not shake. Vigorous mixing can cause hemolysis.
12. As soon as the last tube is filled and mixed, remove the needle from the vein. Apply pressure to puncture site with dry, sterile gauze until bleeding stops.
13. Apply bandage.
14. After venipuncture, the top of the stopper may contain residual blood. Proper precautions should be taken when handling tubes to avoid contact with this blood. Any needle holder that becomes contaminated with this blood should be considered hazardous.
15. After collection, dispose of the needle using and appropriate disposal device. Do not shield.
16. Store tube uptight at room temperature until centrifugation. For best results, blood samples should be centrifuged within 2 h after collection.

3.8 Isolation of Human Peripheral Blood Mononuclear Cells and In Vitro Culture

1. Turn on the Biological Safety Cabinet and UV lamp for 15 min then switch to the visible light.
2. Centrifuge tube/blood samples at room temperature (18–25 °C) in a horizontal rotor (swing-out head) for a minimum of 30 min at 1500 × g (*see* **Note 20**).
3. After centrifugation, peripheral blood mononuclear cells (PBMCs) and platelets will be in a whitish layer just under the plasma layer. Using serological pipette remove approximately half of the plasma without disturbing the cell layer.
4. Collect cell layer with a sterile serological pipette and transfer aseptically into sterile 50 mL size conical centrifuge tube with cap. Collection of cells immediately following centrifugation will yield best results (*see* **Note 21**).
5. Resuspend the PBMCs in RPMI-1640 complete medium to bring volume to 5 mL.
6. Using serological pipette mix the cells up and down for five times.
7. Transfer 100 μL from the cell suspension into microtube and add 300 μL of Trypan Blue to stain cells.
8. Mix well the cell dilution with a pipette (up and down).

9. Transfer 10 μL of the cell dilution into the hemocytometer. Do not overfill. Allow the cell suspension to settle for at least 10 s before counting (*see* **Note 22**). Assess if the cells are evenly distributed among the squares. To get an equal cell distribution, mix cell suspension before adding the stain and again just before loading the hemocytometer.
10. Use a handheld counter to record cell counts. Count viable cells in four large squares. Observe the cell color to differentiate dead from viable cells. Blue cells are dead and clear are alive. Count only live cells.

    To calculate the cell number per mL

    Viable cell/mL = (total number of viable cells/squares counted) $\times 10^4 \times$ dilution factor).

    Example:

    Square 1: counted 50 cells.

    Square 2: counted 53 cells.

    Square 3: counted 49 cells.

    Square 4: counted 45 cells.

    Dilution factor: 4 (Remember, for the dilution preparation we added 300 μL trypan blue to the 100 μL cell suspension).

    Viable cell/mL = $[(50 + 53 + 49 + 45)/4] \times 10^4 \times 4 = 197 \times 10^4$.

    To calculate the total viable cells

    Total viable cells = viable cell/mL × volume of original cell suspension.

    Example:

    Viable cell/mL = 197×10^4.

    Cell suspension volume = 5 mL.

    Total viable cells = $197 \times 10^4 \times 5$ mL = 9.85×10^6.

11. Centrifuge the cell suspension in the 50 mL conical tube for 15 min at 300 × *g* at 20 °C.
12. Aspirate as much supernatant as possible without disturbing the cell pellet.
13. Resuspend the cell pellet by tapping tube with the index finger. DO NOT PIPET OR VORTEX to avoid cell damage.
14. Add 1× PBS to bring volume to 5 mL. Cap tube. Mix cells by inverting tube two times.
15. Centrifuge for 10 min at 300 × *g* at 20 °C (*see* **Notes 23** and **24**).
16. Aspirate as much supernatant as possible without disturbing cell pellet.

17. Resuspend the cell pellet at the desired cell density (cell/mL) with RPMI-1640 complete medium and mix well using a serological pipette (up and down).

    To calculate the volume of RPMI-1640 complete medium to resuspend the cell at a final density

    Volume RPMI = Total cell viable/Cell suspension.

    Example: To prepare a cell suspension at density of 2×10^6 cell.

    Viable cells = 2.45×10^6 cell.

    Cell suspension volume = 18 mL.

    Total cell viable = 2.45×10^6 cell × 18 mL = 44.1×10^6 cell/mL.

    Volume of RPMI = (44.1×10^6 cell/mL)/(2×10^6 cells) = 22 mL RPMI-1640 complete.

3.9 Human-Derived Macrophage and Fatty Acid-Binding Protein Treatment

1. After assessing the cell viability by trypan blue, culture PBMCs in 6-well plates at a density of 4×10^6 cells/well in supplemented RPMI-1640 complete (*see* **Note 25**).
   Example:

   Viable cells = 2.45×10^6 cell.

   Cell suspension volume = 18 mL.

   Total cell viable = (2.45×10^6 cell) (18 mL) = 44.1×10^6 cell/mL.

   Volume of RPMI = *(44.1×10^6 cell/mL)* / (2×10^6 cell)= 22 mL RPMI-1640 complete

   For each well add 2 mL of cell suspension at a density of 2×10^6 cell to obtain a final density of 4×10^6 cells/well.

2. Incubate the PBMCs in a wet chamber for 48 h at 37 °C with 5% CO_2.
3. After the incubation, centrifuge the culture plates at 300 × *g* for 10 min at 25 °C.
4. Using sterile serological pipette remove 1 mL of the supernatant and transfer 1 mL of fresh RPMI-1460 complete media final volume 2 mL of RPMI-1640 complete media per well.
5. Incubate the PBMCs in a wet chamber for 48 h at 37 °C with 5% CO_2.
6. After the incubation, centrifuge the culture plate at 300 × *g* for 10 min at 25 °C.
7. Using sterile serological pipette removes 2 mL supernatant and transfers 5 mL of fresh RPMI-1640 complete. Final volume 5 mL of RPMI media.

8. Incubate the PBMCs in a wet chamber for 48 h at 37 °C with 5% CO_2.
9. After the incubation centrifuge the culture plates at 300 × *g* for 10 min at 25 °C.
10. Aspirate the supernatant and transfer 2 mL of RPMI-1640 complete for each well.
11. Incubate the PBMCs in a humidity chamber for 3 h at 37 °C with 5% CO_2.
12. After the incubation treat cells with LPS (60 μg/mL), *Fh*12 (20 μg/mL), or a mix of LPS + *Fh*12 and 1× PBS for a maximum of 48 h (*see* **Note 26**).
13. Incubate for 24 and 48 h at 37 °C with 5% CO_2.
14. After the incubation, collect the supernatant and used for Nitric Oxide (NO) determinations.
15. Adherent cells (mostly monocyte-derived macrophages; MDM) will be washed with cold, sterile 1× PBS (*see* **Note 27**).
16. Add 2 mL sterile 1× PBS into each well and incubate for 2 min at 37 °C with 5% CO_2.
17. After the incubation removes the MDM using a cell scraper and transfer in a sterile microtube for intracellular arginase activity.

3.10 Nitric Oxide Assay

1. Prepare 1 mL of a 100 μM nitrite solution by diluting the provided 0.1 M Nitrite Standard 1:1000 using RPMI-1460 as a diluent.
2. Designate three columns (24 wells) in the 96-well plate for the Nitrite Standard Reference curve. In row B to H dispense 50 μL of RPMI-1460.
3. Add 100 μL of the 100 μM nitrite solution (**step 1**) in row A.
4. Immediately perform six serial twofold dilutions (50 μL/well) in triplicate down the plate to generate the Nitrite Standard Reference Curve (100, 50, 25, 12.5, 6.25, 3.13, and 1.56 μM), discarding 50 μL from the 1.56 μM set of wells. Do not add any nitrite solution to the last well (0 μM). The final volume in each well is 50 μL. The nitrite concentration range is 0–100 μM.
5. Add 50 μL of culture supernatant to well in triplicate.
6. Using a multichannel pipettor, dispense 50 μL of the Sulfanilamide Solution to all experimental sample and standard sample. Incubate for 10 min in the dark at room temperature.
7. Using a multichannel pipettor, dispense 50 μL of the NED solution to all wells and incubate 10 min in the dark at room temperature. A purple/magenta will begin to form immediately.
8. Measure absorbance after 30 min in a plate reader using a 540 nm filter.

9. Nitrite Oxide concentration should be determined by comparison with a standard curve prepared from a stock solution of 100 μM nitrate.

3.11 Arginase Activity

1. After the monocyte-derived macrophages are collected, prepare a cell suspension at a density of 10^6 cells/mL.
2. Monocyte-derived macrophages will be lysated with 100 μL of RIPA buffer supplemented with a protease inhibitor.
3. Incubate on a rocking platform for 1 h.
4. Centrifugate at 20,000 × *g* at 4 °C for 10 min.
5. Using a micropipette, collect the supernatant without disturbing the pellet.
6. Mix 40 μL lysate supernatant with 10 μL of 5× arginine substrate in a clear bottom 96-well reaction plate.
7. To prepare the sample blank control (OD blank), transfer 40 μL sample without 5× buffer into a new well of the reaction plate.
8. For standard background preparation (OD standard background), transfer 50 μL H_2O into a new well of the reaction plate.
9. Transfer 50 μL of 1 mM urea standard (OD standard) into a new well of the reaction plate.
10. Incubate the reaction plate at 37 °C for 2 h.
11. Add 200 μL of urea reagent to all wells (*see* **Note 28**).
12. Add 10 μL 5× substrate buffer to the sample blank control. Tap the plate and mix.
13. Incubated for 60 min at room temperature and read optical density at 430 nm.
14. Determine the urea levels in the samples using the following equation (*see* **Note 29**).

    OD standard − OD water = OD sample − OD water × [Urea Standard] × 50 × $10^3/(40 \times t)$.

    OD standard − OD water = OD sample − OD water × 10.4 (U/L).

3.12 Cell Viability Assay

1. Prepare a PBMC suspension at density 2 × 10^5 cell/mL.
2. Transfer 200 μL of cell suspension in a 96-well flat bottom plate.
3. Incubate the PBMC for 6 days in a humidity chamber at 37 °C with 5% CO_2. The RPMI medium will be changed every 2 days. Centrifuge the culture plates at 300 × *g* for 10 min at 25 °C and aspirate the supernatant. Transfer 200 μL of RPMI medium.

4. At day 7, change the RPMI as described above (**step 3**) and incubate for 3 h in a humidity chamber at 37 °C with 5% CO_2.
5. After the incubation period, treat the cells separately with LPS (15 μg/mL), Fh12 (5 μg/mL), or (IL-4 ng/mL) for 24 or 48 h [4]. As a negative control, cells should be treated with 1× PBS under the same conditions.
6. Following the incubation, cell viability was assessed by adding 50 μL XTT labeling reagent.
7. Incubate for 24 and 48 h at 37 °C with 5% CO_2 (*see* **Note 30**).
8. Measure the absorbance at these times using a plate reader at 480 nm with a reference wavelength of 650 nm.

4 Notes

1. Transfer the parasites in a 500 mL beaker with abundant 1× PBS. Using the wooden depressor transfer parasite in a clean 500 mL beaker with 1× PBS. Repeat this step until removing all traces of bile and blood. After completing the wash steps, the parasite will be transported to the laboratory in abundant RPMI-1640 medium. The amount of 1× PBS and RPMI-1640 depends on the number of parasites collected.
2. The process takes 3 h at room temperature and 1 h at 90 °C in a water bath.
3. The separation range of Sephadex G-50 column is 1.5×10^3 to 3×10^4.
4. To calculate the area of your column if the size column is 2.6 cm diameter × 100 cm large. For this, you should use the following formula: $A = \pi r^2$. When diameter = 2.6 cm, radius = 1.3 cm, and π = 3.14, the column area is $(3.14) \times (1.3\ \text{cm})^2 = 53\ \text{cm}^2$.
5. To calculate the volumetric flow rate (mL/h), you should use the following formula: VFr = A × LFr. When $A = 53\ \text{cm}^2$, linear flow rate (LFr) = 5 cm/h, VFr = $53\ \text{cm}^2 \times 5$ cm/h = $265\ \text{cm}^3$/h = 4.42 mL/min. It means that the column should be packed at a VFr of 4.42 mL/min, and this is the equivalent of the linear flow of 5 cm/h. The column should be filled up to the 90 cm height of matrix, if the matrix poured into the column is not enough, then to add an extra amount of gel following the recommendations of **steps 3–5**.
6. Plot absorbance 280 nm against "number of fractions." The peak should be symmetric, indicating that packing of the matrix was homogeneous. However, this parameter should be reevaluated if the long bed of column changes or the matrix suffers drying, and it is repacked.

7. The volume collected from sample application to each peak absorbance value is the V_e of each standard protein. Once all this data is available, calculate the distribution coefficient of each standard protein (K_{av}) of each protein. Plot Kav against log MW.
8. If $V_t = 477.6$ mL, then the minimum volume of antigen loading onto the column will be 9.5 mL and the maximum volume 23.8 mL (to obtain the best separation 9–12 mL of FhWWE is recommended). The FhWWE should be mixed with 3% sucrose to increase the sample density.
9. Once you have found two consecutive equal mA readings, collect the fractions.
10. *Quality Control of Purification Process of Fatty Acid-Binding Protein*: To reproduce the same results during the chromatography process, you must follow the following recommendations:
    (a) Work samples on ice or at 4 °C during all procedures to avoid protein degradation,
    (b) After the first FhWWE separation, the greenish color fraction (fraction 1) and the reddish color fraction (fraction 2) should be collected entirely and pooled individually.
11. Measure the volume of the two fractions. Determine the total protein concentration contained in the two fractions by BCA method and determine the total protein concentration in each section and the FhWEE. For this, use the following equation: Total protein concentration = Protein concentration × volume collected for each fraction. Finally, Sum the total protein content of fraction 1 with a total protein concentration of fraction 2 and calculate the percent of protein recovered, taking as 100% the total amount of protein antigen loaded onto the column. Recover of protein antigen in the range of 90–100% is recommended.
12. During the column activation, it is essential to rinse the wall of the column from the top to the bottom using regeneration buffer. This process may take approximately 60 min for a 1.5 mL column.
13. During the equilibration step, the column wall should be rinsed entirely during this process. This process may take approximately 40 min.
14. To avoid protein degradation collects the protein at 4 °C.
15. The presence of endotoxin will be assessed in protein solution before and after endotoxin removal using Chambrex QCL-1000 Chromogenic LAL Endpoint Assay. The sample will be subject to subsequent endotoxin removal processes

until the levels of endotoxins were lower or equal than RPMI-1640 medium, which is considered endotoxin free. If the final endotoxin levels are above the desired endotoxin levels, repeat the procedure by reloading the sample to the regenerate column.

16. Endotoxin Removal Consideration: Endotoxin removal can be performed in solution from pH 5–10. The highest removal occurs at pH 7–8. If the removal efficiency is low; increase the contact time by decreasing the flow rate or consider adjusting the pH to pH = 7–8. Do not reuse the column for a sample containing different target molecules.
17. Each series of determinations must include a blank plus the four standards run in duplicate. The blank well contains 50 μL of LAL Reagent Water instead of a sample.
18. It is essential to be consistent with the reagent addition from well to well or row to row and in the rate of pipetting.
19. If no blood flows into the tube or if blood ceases to flow before an adequate sample (approximately 3.0 mL as minimum blood volume for 4 mL draw and about 6.0 mL minimum blood volume for 8 mL draw) is collected, the following steps are suggested to complete satisfactory collection: (a) Confirm correct position of needle cannula in the vein. (b) a multiple sample needle is being used to remove the tube and place a new tube into the holder. (c) If the second tube does not draw, remove the needle and discard at the appropriate disposal device. Do not reshield. Repeat procedure from **step 1**. When using a blood collection set, a reduced draw of approximately 0.5 mL will occur on the first tube. This reduced draw is due to the trapped air in the blood collection set tubing, which enters the first tube.
20. Remix the blood sample immediately before centrifugation by gently inverting the tube 8–10 times. Also, check to see that the tube is in proper centrifuge carrier/adapter. Excessive centrifuge speed (over 2000 × *g*) may cause tube breakage and exposure to blood and possibly injury.
21. An alternative procedure for recovering the separated mononuclear cell is to resuspend the cells into the plasma by inverting the unopened vacutainer tube gently 5–10 times. This is the preferred method for storing or transporting the separated sample for up to 24 h after centrifugation.
22. After mixing the cell dilution, count cells immediately. If the solution spreads into the two lateral grooves adjoining the grid table, or if there are bubbles in the solution covering the grid table, clean the hemocytometer and repeat the application.

PBMCs may contain a population of erythrocytes. Use caution when counting cells to distinguish lymphocytes from erythrocytes.

23. Since the principle of separation depends on a density gradient, and the density of components varies with temperature, the temperature of the system should be maintained between 18 and 25 °C during separation.
24. Since the principle of separation depends on the movement of formed elements in the blood through the separation media, the RCF should be set at 1500 × *g* to 1800 × *g*. The time of centrifugation should be a minimum of 20 min. The centrifugation time depends on the specimen sample. Some specimens may require up to 30 min for optimal separation. Centrifugation of the vacutainer tube up to 30 min has the effect of reducing red blood cells contamination of the mononuclear cell fraction. Centrifugation beyond 30 min has a little additional effect. The vacutainer tube may be recentrifuged if the mononuclear layer is not disturbed.
25. For six-well culture plate, you need a minimum volume of 2 mL of cell suspension per well. To perform this experiment, you will need to prepare a minimum of 4 six-well culture plates.
26. In each plate, include LPS and 1× PBS treatment.
27. The remaining adherent cells were highly enriched with macrophages (>95%) as assessed by fluorescence-activated cell-sorting staining with the macrophage marker F4/80 (BD Pharmigen, San Diego, CA).
28. Urea reagent stops arginase reaction.
29. The OD sample, OD blank, OD standard, and OD water are optical density values of the sample, sample blank, standard, and water, respectively. The standard urea concentration is 1 mM and *t* is the reaction time (120 min) and 50 and 40 refers to the reaction and sample volume in microliters, respectively. The Unit definition: one unit of arginase activity converts 1 μmole of L-arginine to ornithine and urea per minute at pH 9.5 and 37 °C per million cells.
30. The assay is based on the cleavage of the yellow tetrazolium salt XXT {(sodium 3′-[1-phenylaminocarbonyl]-3,4-tetrazolium]-bis [4-methoxy-6-nitro]benzene sulfonic acid hydrate)} labeling reagent to form an orange formazan dye by active metabolic cells. This conversion only occurs in viable cells. After the incubation period, the orange formazan solution is formed. The formazan dye is soluble in aqueous solution and is directly quantified using a spectrophotometer.

References

1. Esteves A, Ehrlich R (2006) Invertebrate intracellular fatty acid binding proteins. Comp Biochem Physiol C Toxicol Pharmacol 142:262–274
2. Donnelly S, O'Neill SM, Sekiya M, Mulcahy G, Dalton JP (2005) Thioredoxin peroxidase secreted by *Fasciola hepatica* induces the alternative activation of macrophages. Infect Immun 73:166–173
3. Donnelly S, Stack CM, O'Neill SM, Sayed AA, Williams DL, Dalton JP (2008) Helminth 2-Cys peroxiredoxin drives Th2 responses through a mechanism involving alternatively activated macrophages. FASEB J 22:4022–4032
4. Figueroa-Santiago O, Espino AM (2014) *Fasciola hepatica* fatty acid binding protein induces the alternative activation of human macrophages. Infect Immun 82(12):5005–5012

Chapter 12

Possible Role for Toll-Like Receptors in Interaction of *Fasciola hepatica* Excretory–Secretory Products with Human Monocyte Cell Line

Olgary Figueroa-Santiago and Ana Espino

Abstract

This chapter presents a proteomic approach to purify and identify native excretory–secretory products (ESPs) in the range of >10–30 kDa proteins capable of interacting with toll-like receptors (TLRs). Here we present a protocol to fractionate the total ESPs using an ultrafiltration system to recover ESP proteins >10–30 kDa. The fraction of the proteins >10–30 kDa is purified by ion exchange chromatography (IEC) using a mono Q-column in a fast protein liquid chromatography system (FPLC) to separate its components based on charge. Finally, a screening system is presented using THP1-Blue CD14 cells to investigate whether TLRs could also be targeted by *Fasciola hepatica* ESPs and the interaction with TLR4 using HEK293 Blue-TLR4 cells.

Key words *Fasciola hepatica*, Toll-like receptors, Innate response, Excretory–secretory products, Cell culture, Endotoxin removal, Ion exchange chromatography

1 Introduction

Fasciola hepatica secretes and excretes a large number of antigens to induce an immunosuppressive effect to prevent the innate immune response in the host. By definition, excretory–secretory products (ESPs) are products actively exported through secretory pathways and those that may diffuse or leak from the parasite soma [1]. During migration within its host, flukes continually release ESPs into the host environment and for this reason ESP are considered the first parasite antigens that stimulate the host immune response during infection. The ESPs include a mixture of proteolytic enzymes such as cathepsin-L (Cat-L) and antioxidants such as glutathione S-transferase (GST), thioredoxin peroxidase (TPX)

The original version of this chapter was revised. The correction to this chapter is available at https://doi.org/10.1007/978-1-0716-0475-5_18

Martin Cancela and Gabriela Maggioli (eds.), *Fasciola hepatica: Methods and Protocols*, Methods in Molecular Biology, vol. 2137, https://doi.org/10.1007/978-1-0716-0475-5_12, © Springer Science+Business Media, LLC, part of Springer Nature 2020, Corrected Publication 2020

and fatty acid-binding protein (FABP) [2, 3]. Also, ESPs have shown several potential roles in modulating various characteristics of the host's immune response such as: lowering phagocytes activity and antigen presenting ability of peripheral lymphocytes [4], inhibition of production of nitric oxide (NO) by activated macrophage [5], the induction of alternatives activation of human macrophages [6] and the recruitment of alternative activation of macrophage (AAMΦ) which favors the induction of Th2 response [1, 7–10]. This Th2 response is the ideal immunological environment for parasite survival into the host and ensures transmission. These polarized Th2 responses are only possible with an efficient suppression of Th1 cytokines.

The immunomodulatory activities mediated by the interactions of parasite antigens with several Toll-Like Receptors (TLRs) have been reported in *Schistosoma* and *Fasciola*. In *Schistosomes,* these immunomodulators activities are mediated by the interaction of parasite antigens with TLR2, TLR4, and TLR3 [11–13]. There are only a few studies that have explored the anti-inflammatory properties of *F. hepatica*. These studies have been performed with recombinant forms of only two significant ESPs: CatL1 and GST [14]. It has also been reported the effect of two *F. hepatica* recombinant proteins, CatL1, and GST, on dendritic cells (DCs) from mice and demonstrating that CatL1 and GST can partially activate DCs [13]. Both molecules induce IL-6, IL12p40, and MIP-2 secretion from DCs and enhance CD40 expression via interaction with TLR4 [14]. Recently, we have been reported *Fasciola hepatica* ESPs could activate TLR4, TLR2, TLR8, TLR5, and TLR6, and other ESP components like Fh12 can exert and inhibitory effects on TLRs [14, 15]. In the present chapter protocols related to ESP antigen purification and ESP-TLR interactions are described.

2 Materials

2.1 Preparation of Excretory–Secretory Products

1. Hood.
2. Centrifuge.
3. Sterile wooden depressor.
4. Sterile petri dish.
5. Dissecting microscope.
6. Sterile conical tubes.
7. RPMI-1640 medium.
8. Sterile microtubes.
9. PD-10 column.
10. Column stand.
11. Filter 0.2 μm membrane.

12. Lyophilizer.
13. BCA kit.
14. High flow ultrafiltration system (AMICON).
15. YM-3, YM-100, YM-30 and YM-10 membrane.
16. Sterile Phosphate Buffered Saline (PBS) 1× equilibrated at 37 °C.
17. RPMI-1640 medium supplemented with 25 mM HEPES buffer, 7.5% sodium bicarbonate, 100 μg/mL penicillin, 100 μg/mL streptomycin.
18. Polymyxin B (PMB) column.

2.2 Antigenicity of ESPs

1. 96 wells flat bottom polystyrene plates.
2. Coating buffer: Dissolve in 900 mL dH_2O 1.59 g Na_2CO_3, 2.93 g $NaHCO_3$, 0.20 g sodium azide. Adjust pH 9.6 with HCl and complete volume until 1 L. Store at 4 °C no longer than 1 month.
3. Wash buffer (PBS-T20): 1 L PBS 1×, 0.05% Tween 20.
4. Blocking solution: Dissolve 3 g Albumin in 80 mL PBS-T20. Add PBS-T20 to total volume 100 mL.
5. Substrate buffer: Dissolve in 900 mL dH_2O 1.49 g Na_2HPO_4, 1.02 g citric acid. Adjust pH to 4.5–5 and complete volume until 1 L. Store at 4 °C.
6. Substrate solution: dissolve 10 mg ortho-phenylenediamine hydrochloride (OPD) in 20 mL of substrate buffer and add 10 μL of H_2O_2. Complete volume until 25 mL.
7. Stop solution: 2 M sulfuric acid.

2.3 Ion Exchange Chromatography and Protein Identification

1. Filter 0.45 μm membrane.
2. PD-10 column.
3. Column stand.
4. Lyophilizer.
5. BCA kit.
6. Mono Q 5/50 GL column and the AKTA FPLC (Amersham-Bioscience).
7. 10 mM Tris–HCl pH 8.0: Dissolve 1.21 g Tris in 900 mL dH_2O. Adjust pH with HCl and complete volume until 1 L. Filter with 0.45 μm membrane. Store at 4 °C (*see* **Note 1**).
8. Elution buffer: 10 mM Tris–HCl pH 8.0, 1 M NaCl. Dissolve 29.22 g NaCl in 500 mL of 10 mM Tris–HCl pH 8.0. Complete volume until 1 L. Filter with 0.45 μm. Store at 4 °C.

2.4 Screening System Using THP1-Blue CD14 Cell Treats with *Fasciola hepatica* ESPs Fractionated with an Ultrafiltration System

1. THP1-Blue™-CD14 cells.
2. Endotoxin-free flat-bottomed 96 well plates.
3. Wet chamber.
4. TLR 1–9 agonists kit by Invitrogen.
5. *Listeria monocytogenes* (HKLM): 5×10^7 cells/mL.
6. Lipopolysaccharide (LPS): 1 μg/mL, TLR-4 ligand.
7. Flagellin (FLA): 100 ng/mL.
8. Thiazoloquinoline agonist (CL075): 0.5 μg/mL; TLR 8/7 ligand.
9. Microplate reader.
10. Endotoxin water.
11. Spectrophotometer or filter photometer with 665 nm.
12. Oxidized 1-palmitoyl-2-arachidonyl-*sn*-glycero-3-phosphorylcholine (OxPAPC): TLR2 and TLR4 inhibitor.
13. Polymyxin B (PMB): TLR4 inhibitor.
14. Chloroquine (Chlor): TLR-3, 7, 8, 9 inhibitor.
15. RPMI-1640 (Invitrogen) supplemented with 2 mM L-glutamine, 1.5 g/L sodium bicarbonate, 4.5 g/L glucose, 10 mM HEPES, 1.0 mM sodium pyruvate, 10% inactivated fetal bovine serum (FBS). Add 50 U/mL Penicillin–Streptomycin, Blasticidin 10 μg/mL, and Zeocin 200 μg/mL.
16. Quanti-blue reagent: dissolve one pouch of Quanti-blue (QB) in 100 mL sterile dH_2O. Filter with the 0.2 μm membrane. Warm the QB reagent at 37 °C before use (*see* **Note 2**).

2.5 NF-κB Activation in TLR4-Transfected HEK Cells Treated with *Fasciola hepatica* ESPs Fractionated with IEC Chromatography

1. HEK-Blue-hTLR4 cell line by Invitrogen.
2. DMEM media supplemented with 4.5 g/L glucose, 10% inactivated fetal bovine serum (FBS), 50 U/mL penicillin, 50 μg/mL streptomycin, 100 μg/mL Normocin, and 2 mM L-glutamine.
3. Lipopolysaccharide (LPS): 1 μg/mL, TLR-4 ligand.
4. PMB (TLR4 inhibitor).
5. Endotoxin-free flat-bottomed 96-well plates.
6. Wet chamber.
7. Microplate reader.
8. Quanty Blue.
9. Endotoxin water.
10. Spectrophotometer or filter photometer with 665 nm.

3 Methods

3.1 Preparation of Excretory–Secretory Products (ESP)

1. Adult flukes were collected from bile ducts of cattle sacrificed at a local slaughterhouse and repeatedly washed with 1× PBS to remove all traces of blood and bile (*see* Chapter 11: *Purification of native Fasciola hepatica Fatty Acid Binding Protein and the induction of Alternative Activation of human Macrophages*, **step 1**, Subheading 3.1).
2. Once at the laboratory, transfer flukes into a beaker and wash with PBS 1×. Repeat this step until the parasites are as clean as possible (*see* **Note 3**).
3. Using a wood depressor transfer flukes into a petri dish with PBS 1×.
4. Using a dissecting microscope, evaluate the fluke motility and structures (*see* **Note 4**).
5. Flukes will be individually incubated overnight in conical tubes under sterile conditions at 37 °C in RPMI-1640 medium at a ratio of one fluke in 5 mL of medium (*see* **Note 5**).
6. After incubation, flukes were removed, and the culture medium from each tube was pooled and centrifuged (1500 × *g* for 10 min at 4 °C) to remove parasite eggs (*see* **Note 6**).
7. Approximately 1 L of the supernatant containing the parasite ESPs will be collected.
8. One hundred-mL (total ESPs) will be concentrated up to a volume of 5 mL using a high flow ultrafiltration system (AMICON), using a YM-3 membrane, which retains in its surface proteins ≥3 kDa (*see* **Note 7**).
9. The remaining 900 mL will be submitted to a sequential fractionation by AMICON to separate proteins in different ranges of molecular weight (MW) (*see* **Note 8**).
10. To separate proteins of large MW, 900 mL of ESPs will be passed through a YM-100 membrane.
11. Molecules retained on the surface of the membrane (ESPs > 100 kDa) will be suspended in 2 mL of sterile PBS 1× and centrifuge during 20 min at 1500 × *g* to eliminate gross unsolved particles and sterilized by filtration using Millipore filter 0.2 μm.
12. Proteins that passed through YM-100 membrane (flow through) will be subsequently passed through a YM-30 membrane retaining molecules ≥30 kDa. Proteins retained on YM-30 membrane (ESPs > 30–60 kDa) will be suspended in 2 mL of PBS, centrifuge and sterilize by filtration as described above.

13. The flow-through will be ultrafilter using a YM-10 membrane keeping molecules ≥10 kDa. Proteins retained on the surface were designated as ESPs >10–30 kDa.
14. ESPs >10–30 kDa will also be centrifuged and filtered as described above. The flow-through of the YM-10 membrane was discharged.
15. Using PD-10 columns desalt total or fractionated ESPs against PBS 1×.
16. Endotoxin will be removed from all samples using PMB column, and the presence of endotoxins will be asses before and after removing endotoxin (*see* Chapter 11: *Purification of native Fasciola hepatica Fatty Acid Binding Protein and the induction of Alternative Activation of human Macrophages*, Subheadings 2.5, 2.6, 3.5 and 3.6).
17. Proteins will be lyophilized and resuspended in 2 mL of sterile dH_2O (*see* **Note 9**).
18. Determine the protein concentration using the BCA method.
19. Prepare aliquots of 500 μL and store at −20 °C until use.
20. Analyzed the total and the fractionated ESPs by 12% SDS-PAGE.

3.2 Antigenicity of ESPs

1. Coat the wells with 35 μg/mL of ESPs in coating buffer and incubate the plate overnight at 4 °C.
2. Wash the plate three times with PBS-T20.
3. Block wells with 300 μL/well of 3% albumin in PBS-T20 and incubate 1 h at 37 °C into a wet chamber.
4. Discard the blocking solution.
5. Add 100 μL/well of the serum samples diluted 1:200 in PBS-T20 and incubated for 1 h at 37 °C in a wet chamber.
6. Wash the plate three times with PBS-T20.
7. Add 100 μL/well of an anti-rabbit IgG conjugated with peroxidase (diluted 1:5000 in PBS-T20) and incubate 1 h at 37 °C in a wet chamber.
8. Wash the plate three times with PBS-T20.
9. Add 100 μL/well of substrate solution and incubate 30 min at room temperature in the dark.
10. Stop the reaction by adding 50 μL/well of 12.5% sulfuric acid.
11. Read the plate at 492 nm.

3.3 Ion-Exchange Chromatography

1. Using a dialysis membrane, dialyze exhaustedly the total or fractionated ESPs against 5 L of 10 mM Tris–HCl pH 8.0 overnight at 4 °C.
2. Filter the sample with a 0.45 μM membrane.
3. Using a Mono Q 5/50 GL column and the AKTA FPLC fractionate the sample.
4. Elute with 10 mM Tris–HCl pH 8.0, 1 M NaCl, using a step-wise protocol with a flow rate of 1 mL/min.
5. Collect the void fractions as well as the eluent (IEC fractions) with detectable UV absorption peaks (*see* **Note 10**).
6. Using a PD-10 column desalt samples.
7. Remove Endotoxin and detect endotoxin levels.
8. Using a lyophilizer concentrate samples and reconstituted with sterile PBS 1×.
9. Determine the protein concentration using the BCA method.
10. Analyzed IEC fractions by 12% SDS-PAGE (*see* **Note 11**).
11. Prepare aliquots and store at −20 °C until use.

3.4 Screening System Using THP1-Blue CD14 Cell Treats with Fasciola hepatica ESPs Fractionated with an Ultrafiltration System

1. Seed THP1-Blue CD14 cells in a sterile endotoxin-free flat-bottomed 96 well plates (Costar) within RPMI-1640 at a density of 2 × 10^6 cells/mL (*see* Chapter 11: *Purification of native Fasciola hepatica Fatty Acid Binding Protein and the induction of Alternative Activation of human Macrophages*, **steps 7–17**, Subheading 3.8).
2. Incubate in a wet chamber at 37 °C, 5% CO_2 for 3 h.
3. After incubation, treat the cells with 15 μg/mL of the ESPs antigens or TLRs fractions. To test TLR agonists use the following agonist and concentrations: HKLM, LPS (1 μg/mL), FLA (100 ng/mL), and CL075 (0.5 μg/mL) (*see* **Note 12**).
4. Incubate in a wet chamber at 37 °C, 5% CO_2 for 19 h.
5. After incubation, add 150 μL of the QB reagent into each well (*see* **Note 13**).
6. Incubate for 7 h in a wet chamber and measure the absorbance at a wavelength of 655 nm (A_{655}).
7. For the inhibition experiments, treat first the cell with each TLRs antagonist at the following concentration: OxPAPC (30 μg), PMB (100 μg) or Chlor (100 μM) (*see* **Note 14**).
8. Incubate for 30 min in a wet chamber at 37 °C, 5% CO_2.
9. After incubation, treat the cells with ESP-fractions or ligands.
10. Incubation for 19 h in a wet chamber and add the QB reagent as described above.

11. To determine the reduction in the absorbance value, the criteria of specific activation for a given TLR should be used, and calculated using the following formula: $R(\%) = (C - E)/C \times 100$. In the equation, C represents the mean A_{665} of three replicates obtained when cells are stimulated with ESPs, or ligands and E represent the mean A_{665} value obtained when cells were first exposed to the TLR-inhibitors and then stimulated with the ESPs or ligands (*see* **Note 15**).

3.5 NF-κB Activation in TLR4-Transfected HEK Cells Treated with *Fasciola hepatica* ESPs Fractionated with IEC Chromatography

1. Seed TLR4-transfected HEK cells at a density of 2.52×10^4 cells/well in a sterile 96-well flat-bottom plates (*see* **step 1**, Subheading 3.4).
2. Incubate in a wet chamber at 37 °C, 5% CO_2 for 3 h.
3. After the incubation, treat the cells with IEC-fraction (15 μg/mL) or LPS (1 μg/mL).
4. Incubated in a wet chamber at 37 °C, 5% CO_2 for 19 h.
5. For the inhibition experiments, cells were cultured with PMB (100 mM) and incubated for 30 min in a wet chamber.
6. After the incubation, treat the cells with LPS (1 μg/mL) or IEC-fraction. The percent of reduction of the absorbance values was calculated as described above.

4 Notes

1. 10 mM Tris–HCl pH 8.0 will be used for dialysis and as equilibration buffer.
2. Some FBS may contain alkaline phosphatase that can interfere with SEAP quantification. It is recommended to test the culture media supplemented with FBS to evaluate the presence of alkaline phosphatase.
3. At the laboratory, the procedure will be performed under the sterile condition and work at hood as the procedure so allow.
4. For culture parasites only use motile and undamaged parasites.
5. If the parasites are collected before noon, you should culture the flukes in 3 mL of media for 4 h. After the incubation, collect the media and add 5 mL of media and then incubated overnight at 37 °C.
6. Before centrifugation, evaluate the parasite's motility and integrity. Discard culture with dead parasites to avoid the presence of the protease.
7. During the concentration process covers the AMICON system with ice pad to avoid protein degradation.

8. During all the sequential fractionation process cover the AMICON system and the collecting tubes with ice to avoid protein degradation.
9. If you do not have a lyophilizer, you can concentrate the sample using a YM3 membrane.
10. The eluent contains the ion exchange chromatography fraction (IEC fractions). Perform three consecutive runs and collect IEC fractions from runs 1, 2 and 3 with identical elution volume. Pool the samples obtained in each of the runs.
11. Major protein bands were manually excised from the electrophoresis gel, analyzed by MALDI-MS/MS and identify by comparison with molecules deposited at Swiss-Prot and NCBInr databases using the MASCOT search engine (Matrix-Science, London, UK) [6].
12. Cells stimulated with TLR-agonists at the concentration suggested by the manufacturer will be used as positive activation control. Wells treated with sterile PBS 1× as a negative control.
13. Another method is to transfer 20 μL of the supernatant from each well into a clean 96-well microplate and mixed with 150 μL of the QB reagent.
14. As a positive control, cells will be stimulated with specific TLR ligands. For the negative control, cells were exposed to a specific TLR inhibitor.
15. All determinations will be in triplicate, and each experiment will be repeated three times. The results are expressed as the mean A_{655} values ± SD for each determination. Statistical significance among different experimental determinations will be performed using unpaired Student *t*-test using GraphPad Prism software (Prism 6) a *p*-value ≤ 0.05.

References

1. Hewitson JP, Grainger JR, Maizels RM (2009) Helminth immunoregulation: the role of parasite secreted proteins in modulating host immunity. Mol Biochem Parasitol 167(1):1–11
2. Hacariz O, Sayers G, Baykal AT (2012) A proteomic approach to investigate the distribution and abundance of surface and internal *Fasciola hepatica* proteins during the chronic stage of natural liver fluke infection in cattle. J Proteome Res 11:3592–3604
3. Jefferies JR, Campbell AM, van Rossum AJ, Barrett J, Brophy PM (2001) Proteomic analysis of *Fasciola hepatica* excretory-secretory products. Proteomics 1:1128–1132
4. Cervi L, Rubinstein H, Masih D (1996) Involvement of excretion-secretion products from *Fasciola hepatica* inducing suppression of the cellular immune responses. Vet Parasitol 61:97–111
5. Cervi L, Rossi G, Masih DT (1999) Potential role for excretory-secretory forms of glutathione-S-transferase (GST) in *Fasciola hepatica*. Parasitology 119(6):627–633
6. Figueroa-Santiago O, Espino AM (2014) *Fasciola hepatica* fatty acid binding protein induces the alternative activation of human macrophages. Infect Immun 82(12):5005–5012
7. Donnelly S, O'Neill SM, Sekiya M, Mulcahy G, Dalton JP (2005) Thioredoxin peroxidase secreted by *Fasciola hepatica* induces the alternative activation of macrophages. Infect Immun 73:166–173

8. Flynn RJ, Mannion C, Golden O, Hacariz O, Mulcahy G (2007) Experimental *Fasciola hepatica* infection alters responses to tests used for diagnosis of bovine tuberculosis. Infect Immun 75(3):1373–1381
9. Piedrafita D, Estuningsih E, Pleasance J, Prowse R, Raadsma HW, Meeusen ENT, Spithill TW (2007) Peritoneal lavage cells of Indonesian thin-tail sheep mediate antibody-dependent superoxide radical cytotoxicity in vitro against newly Excysted juvenile *Fasciola gigantica* but not juvenile *Fasciola hepatica*. Infect Immun 75(4):1954–1963
10. Zhang W, Moreau E, Peigne F, Huang W, Chauvin A (2005) Comparison of modulation of sheep, mouse and buffalo lymphocyte responses by *Fasciola hepatica* and *Fasciola gigantica* excretory-secretory products. Parasitol Res 95(5):333–338
11. Aksoy E, Zouain CS, Vanhoutte F, Fontaine J, Pavelka N, Thieblemont N, Willems F, Ricciardi-Castagnoli P, Goldman M, Capron M et al (2005) Double-stranded RNAs from the helminth parasite Schistosoma activate TLR3 in dendritic cells. J Biol Chem 280 (1):277–283
12. Thomas PG, Carter MR, Atochina O, Da'Dara AA, Piskorska D, McGuire E, Harn DA (2003) Maturation of dendritic cell 2 phenotype by a helminth glycan uses a toll-like receptor 4-dependent mechanism. J Immunol 171:5837–5841
13. Dowling DJ, Hamilton CM, Donnelly S, La Course J, Brophy PM, Dalton J, O'Neill SM (2010) Major secretory antigens of the helminth *Fasciola hepatica* activate a suppressive dendritic cell phenotype that attenuates Th17 cells but fails to activate Th2 immune responses. Infect Immun 78:793–801
14. Figueroa-Santiago O, Espino AM (2017) *Fasciola hepatica* ESPs could indistinctly activate or block multiple toll-like receptors in a human monocyte cell line. Ann Clin Pathol 5(3):1112
15. Figueroa-Santiago O, Delgado B, Espino AM (2011) *Fasciola hepatica* saposin-like-2 - protein-based ELISA for the serodiagnosis of chronic human fascioliasis. Diagn Microbiol Infect Dis 70(3):355–361

Chapter 13

Evaluation of the Immune Regulatory Properties of Dendritic Cells During *Fasciola hepatica* Infection

Teresa Freire

Abstract

Dendritic cells (DCs) are potent antigen-presenting cells that possess the ability to stimulate *naïve* T cells, initiating the adaptive immune response. Ex vivo DC cultures are useful to evaluate how helminths regulate DC maturation and stimulatory activity. Here, we describe how to isolate CD11c$^+$ from *F. hepatica*-infected mice to evaluate their activation state, cytokine production and regulatory function in an allogeneic T cell assay.

Key words *Fasciola hepatica*, Immune regulation, Dendritic cell, Maturation, CD11c, Cytokines, Mixed lymphocyte reaction, T cell proliferation

1 Introduction

To allow long survival in the host, *Fasciola hepatica* evades host immunity by altering dendritic cell (DC) maturation and function [1–5], resulting in an adaptive immune response known as modified Th2/Treg polarization. DCs are potent antigen presenting cells that possess the ability to stimulate naive T cells. In response to infectious agents DCs undergo a maturation process during which they migrate to secondary lymphoid organs where they present captured antigens to *naïve* T cells, and trigger specific immunity. This process is associated to an upregulation of the expression of MHC molecules, adhesion molecules, and costimulatory molecules (CD40, CD80, or CD86) as well as a downregulation of their endocytic capacity [6, 7]. However, in the presence of *F. hepatica*-derived molecules or during the parasite infection, mature DCs express reduced levels of costimulatory markers and MHC class II molecules, as compared to DCs matured with Toll-like receptor (TLR), a state that is been known as semi-maturation [8, 9]. These DCs are not capable of producing high levels of pro-inflammatory cytokines (IL-12, IL-6, or TNFα) but they may produce anti-inflammatory cytokines such as IL-10 or TGFβ [3, 5, 8, 9]. Finally,

Martin Cancela and Gabriela Maggioli (eds.), *Fasciola hepatica: Methods and Protocols*, Methods in Molecular Biology, vol. 2137, https://doi.org/10.1007/978-1-0716-0475-5_13, © Springer Science+Business Media, LLC, part of Springer Nature 2020

F. hepatica components also modulate the stimulatory activity or function of DCs. Indeed, our group and others have independently demonstrated that during the infection *F. hepatica* inhibits or decrease DC activation, which results in the induction of a tolerogenic phenotype [3–5, 10–13]. Thus, it has been hypothesized that *F. hepatica* regulates DC function and fate as a mean to control its pathogenesis and survival in the infected hosts.

In this context, ex vivo DC cultures are useful to evaluate how the parasite can modulate or inhibit their activation and stimulatory function during the infection. In this chapter, we describe how to isolate DCs from *F. hepatica*-infected mice and report how to culture them in order to evaluate their activation state, cytokine production and regulatory function in an allogeneic T cell assay. To this end, we purify CD11c+ cells from the peritoneum of infected animals and evaluate their maturation and influence on allogeneic CD4 T cell proliferation (Fig. 1). The mixed lymphocyte reaction is a standard way to evaluate the regulatory capacity of T cells that proliferate upon recognition of allogeneic MHC molecules as foreign and divide.

2 Materials

2.1 Infection

1. Female BALB/c mice (6–8 weeks old).
2. *F. hepatica* metacercariae.
3. Low-dose anesthesia: ketamine (20 mg/kg) and xylazine (1 mg/kg) in sterile PBS (optional).
4. Mouse oral gavage.

2.2 Purification of CD11c+ Cells

1. Sterile phosphate buffered saline pH 7.4 (PBS).
2. Sterile ACK erythrocyte lysing buffer. 1.5 M NH_4Cl, 100 mM $KHCO_3$, 10 mM EDTA, pH 7.2. Mix all reagents in distilled water, adjust the pH and filter sterile. Keep a 10× stock at 4 °C up to 6 months.
3. Anti-mouse CD11c magnetic beads/particles and magnet.
4. Staining buffer for FACS. 2% FBS in PBS, sterile (*see* **Note 1**).
5. Complete culture medium. Sterile RPMI-1640 with glutamine supplemented with 10% heat-inactivated fetal bovine serum (FBS), 50 μM 2-mercaptoethanol, 100 U/ml penicillin, and 100 mg/ml streptomycin.
6. Sterile conic tubes (15 and 50 ml).
7. Hemacytometer.
8. Sterile flow hood.
9. Centrifuge Sorvall ST-16.

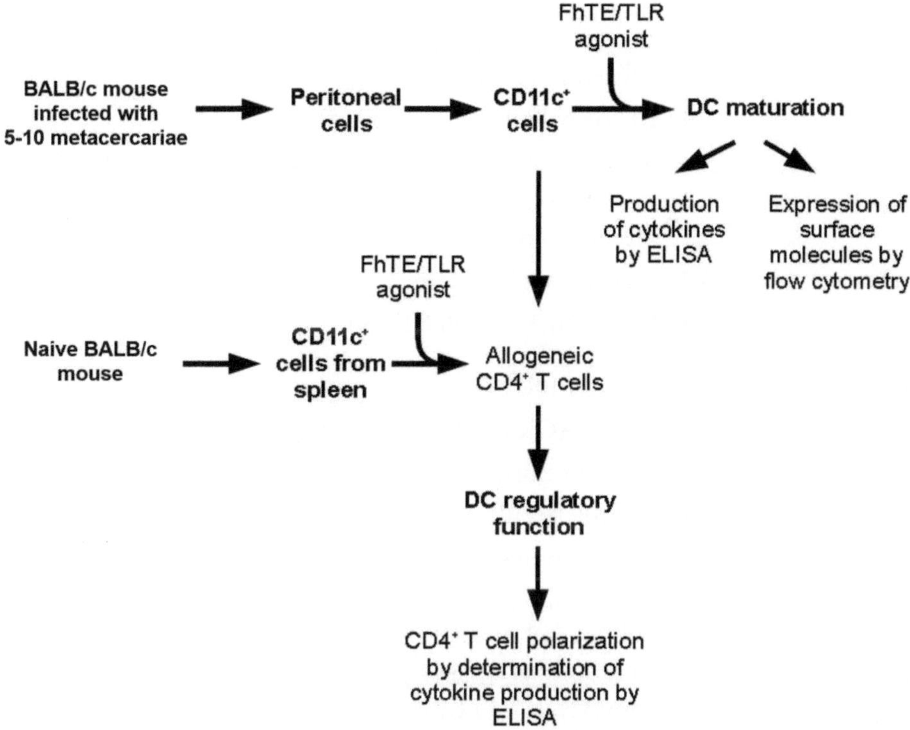

Fig. 1 Analysis of DC maturation and stimulatory function. CD11c$^+$ cells purified from the peritoneum of *F. hepatica*-infected animals are stimulated with a parasite total extract (FhTE) in the presence or absence of maturation stimuli such as a TLR ligand. Then, maturation is evaluated through the quantification of produced cytokines and expression of MHC class II and costimulatory molecules. Alternatively, FhTE-stimulated DCs can be cocultured with allogeneic CD4$^+$ T cells to evaluate the regulatory function through the evaluation of cytokine production

2.3 Evaluation of DC Maturation

1. Sterile PBS.
2. Sterile ACK erythrocyte lysing buffer.
3. Staining buffer for FACS.
4. Complete culture medium.
5. 96-well polystyrene conical bottom microwell plates.
6. 96-well polystyrene flat bottom microwell culture plates.
7. *F. hepatica* total extract (FhTE). Wash live adult worms in PBS. Then, mechanically disrupt the worms with a tissue grinder Potter-Elvehjem and sonicate. Centrifuge at 40,000 × *g* for 60 min and collect the supernatant. Resuspend on PBS containing a cocktail of protein inhibitors and dialyze against PBS for 24 h at 4 °C with a 3 kDa cutoff dialysis membrane. To remove endotoxin contamination, apply FhTE to a column containing endotoxin-removing gel. Determine protein concentration and prepare a sterile solution at 200 μg/ml diluted in complete culture medium.

8. TLR agonist, that is, lipopolysaccharide (LPS, TLR4 ligand) or Pam2CysK4 (TLR2 ligand). Prepare a solution at 4 μg/ml diluted in complete culture medium.
9. 1% formaldehyde in sterile PBS.
10. Specific antibodies for determination of IL-12p40, IL-12p70, TNFα, IL-10, TGFβ by sandwich ELISA.
11. Hemacytometer.
12. Sterile flow hood.
13. Centrifuge Sorvall ST-16.

2.4 Mixed Lymphocyte Reaction

1. Sterile PBS.
2. Sterile ACK erythrocyte lysing buffer.
3. Staining buffer for FACS.
4. Complete culture medium.
5. 96-well polystyrene flat bottom microwell culture plates.
6. *F. hepatica* total extract (FhTE). Prepare a sterile solution at 200 μg/ml diluted in complete culture medium.
7. TLR agonist, that is, lipopolysaccharide (LPS, TLR4 ligand) or Pam2CysK4 (TLR2 ligand). Prepare a solution at 4 μg/ml diluted in complete culture medium.
8. Spleen from a BALB/c *naïve* mouse for splenic CD11c+ cell purification.
9. 100 U/ml DNase-I in complete culture medium.
10. 4000 U/ml Collagenase-D in complete culture medium.
11. Mouse CD4 T cell Negative selection kit and magnet. Prepare obtained cell suspension at 1 × 10^7 cells/ml diluted in complete culture medium.
12. Specific antibodies for determination of IFNγ, IL-4, IL-10, and IL-17 by sandwich ELISA.
13. Hemacytometer.
14. Sterile flow hood.
15. Centrifuge Sorvall ST-16.

3 Methods

3.1 Obtaining of Peritoneal CD11c$^+$ Cells from F. hepatica-Infected Mice

1. Infect BALB/c mice orally with a mouse oral gavage (5–10 metacercariae per mouse in 100–200 μl PBS) (*see* **Note 2**). After the desired period of time (i.e., 1, 2, or 3 weeks postinfection), gently sacrifice the animals by dislocation. Sanitize the animals by pouring alcohol 70% on the peritoneum. Work in a sterile flow hood in order to ensure cell sterility. Inject i.p. with

a 21G syringe 5 ml of cold sterile PBS (*see* **Note 3**). Inject additional 5 ml of cold PBS. Gently massage the peritoneum. Then, perform a small cut in the peritoneum and quickly transfer the liquid to a sterile conic 50 ml tube.

2. Examine the liver to determine hepatic lesions.
3. Centrifuge the peritoneal cells at 320 × *g* for 7 min at 4 °C. Discard the supernatant. Resuspend cells in 1 ml of ACK buffer and incubate for 1 min to lyse the erythrocytes (*see* **Note 4**).
4. Centrifuge the cells at 320 × *g* for 7 min at 4 °C and discard the supernatant and wash the cells twice with 10 ml of PBS.
5. Resuspend cells in 1 ml of complete culture medium. Proceed to cell counting with trypan blue exclusion to determine cell viability *using* a hemacytometer (*see* **Note 5**).
6. Purify CD11c$^+$ cells by positive selection by incubating with anti-CD11c magnetic beads according to the instructions of the selected manufacturer. Obtained purity should be higher than 90% (*see* **Note 6**). These cells can be used either to (1) quantify the production of cytokines and expression of surface costimulatory molecules upon in vitro stimulation (Fig. 2) or (2) evaluate their regulatory function by assessing the inhibition of and allogeneic CD4 T cell proliferation assay (Fig. 3).

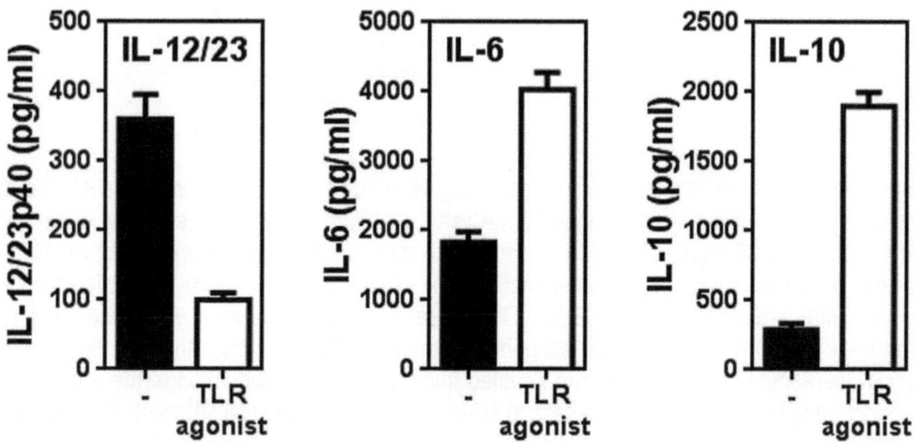

Fig. 2 Cytokine production by peritoneal DCs from *F. hepatica*-infected mice. CD11c$^+$ purified cells from the peritoneal cavity of infected mice were stimulated with the TLR4 agonist LPS overnight at 37 °C. Then, IL-12/23p40, IL-6 and IL-10 production by CD11c$^+$ cells was evaluated on culture supernatants by ELISA. Peritoneal CD11c$^+$ cells from *F. hepatica*-infected mice produced high levels of IL-6 and IL-10 and low levels of the proinflammatory cytokine IL-12/23p40 upon TLR4 stimulation. Asterisks indicate statistically significant differences ($*p < 0.001$)

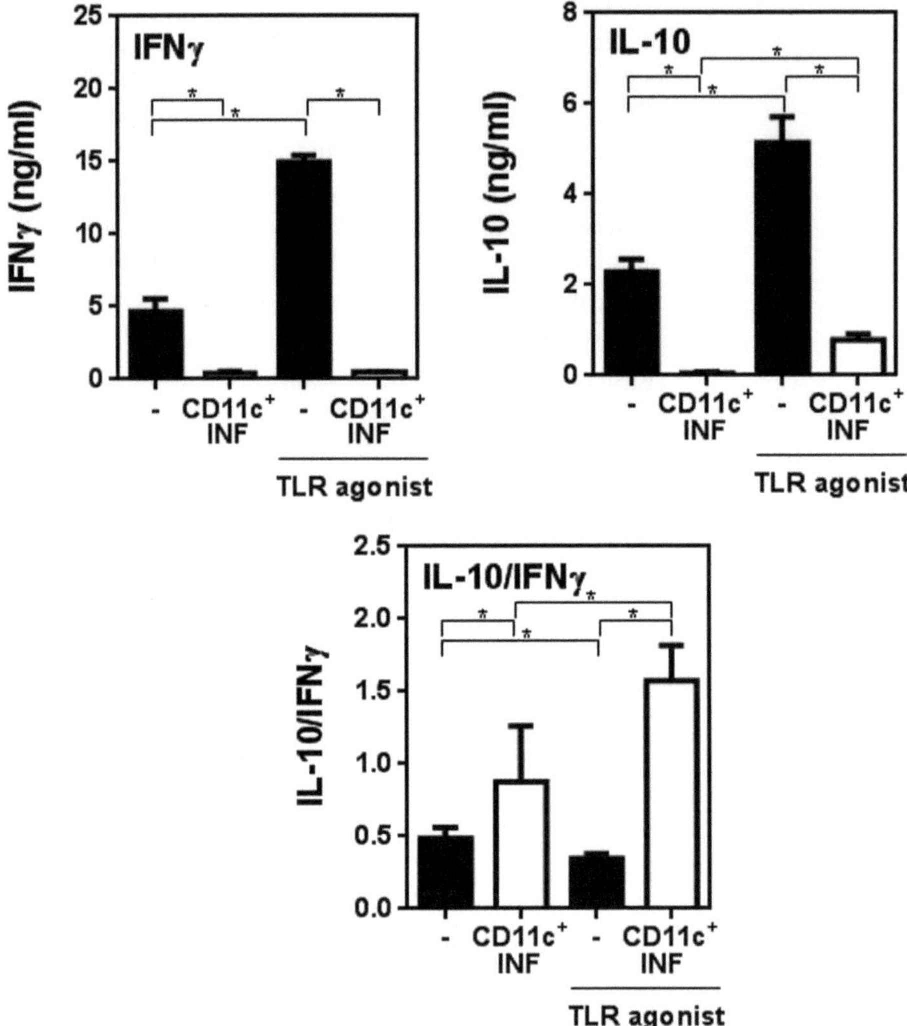

Fig. 3 Regulatory function of DCs in an allogeneic assay. Splenic $CD11c^+$ cells from *naïve* BALB/c mice were stimulated with or without the TLR2 agonist Pam2CysK4 overnight at 37 °C. Then, cells were washed and incubated with purified C57BL/6 *naïve* $CD4^+$ T cells for 5 days at 37 °C in the presence of $CD11c^+$ purified cells from the peritoneal cavity of 3 week-infected mice. Then, IFNγ and IL-10 production by T cells was evaluated on culture supernatants by ELISA. Peritoneal $CD11c^+$ cells from *F. hepatica*-infected mice suppressed IFNγ and IL-10 production by allogeneic stimulated $CD4^+$ T cells. Moreover, in the presence of a TLR agonist, peritoneal $CD11c^+$ cells from infected mice favored the production of the regulatory cytokine IL-10, demonstrating the capacity of these cells to skew $CD4^+$ T allogeneic cell differentiation and suppressing Th1 differentiation. Asterisks indicate statistically significant differences ($*p < 0.001$)

3.2 Evaluation of Cytokine Production and Expression of Surface Molecules by Peritoneal CD11c+ Cells from Infected Animals

1. Gently resuspend the obtained CD11c+ cells in complete culture medium at a concentration of 2.5–5 × 10^6 cells/ml. Dispense 100 μl/well into a 96-well flat bottom microwell culture plate. Incubate in triplicates in presence or absence of activation stimuli (i.e., TLR ligands such as LPS or Pam2CysK4 in 50 μl/well at 4 μg/ml) overnight at 37 °C and 5% CO_2 (*see* **Note 7**). Alternatively, incubate with FhTE (50 μl/well at 200 μg/ml).
2. Collect the cell culture medium and analyze the concentration of pro-inflammatory (IL-12p40, IL-12p70, TNFα) or anti-inflammatory (IL-10, TGFβ) by specific sandwich ELISA (*see* **Note 8**) (Fig. 2).
3. Resuspend the cells in 50 μl FACS staining buffer and place them into a 96-well conical bottom microwell plate. Centrifuge at 560 × *g* for 2 min at 4 °C and discard the supernatant.
4. Incubate the cells with appropriate dilutions of fluorescent conjugated anti-MHC II, -CD40, -CD80, and -CD86 antibodies diluted in FACS staining buffer (50 μl/well) for 15–30 min in the refrigerator.
5. Add 150 μl of FACS staining buffer and centrifuge at 560 × *g* for 2 min at 4 °C. Repeat an additional wash and discard the supernatant (*see* **Note 9**).
6. Incubate cells with 1% formaldehyde in PBS for 15 min in the refrigerator. Wash the cells, discard the supernatant and resuspend them in FACS staining buffer.
7. Analyze in a flow cytometer (*see* **Note 10**).

3.3 Regulatory Function in a Mixed Lymphocyte Reaction

1. Obtain splenic CD11c+ cells from *naïve* BALB/c mice according to the selected manufacturer's instructions (*see* **Note 11**). To this end, collect the spleen, cut it in small pieces and incubate them in complete culture medium supplemented with DNAse I and collagenase D at 37 °C for 1 h. Disaggregate tissue and centrifuge the resuspended cells at 320 × *g* for 7 min at 4 °C. Discard the supernatant.
2. Resuspend cells in 1 ml of ACK buffer and incubate for 1 min to lyse the erythrocytes (*see* **Note 4**). Centrifuge the cells at 320 × *g* for 7 min at 4 °C and discard the supernatant. Wash the cells twice with 10 ml of PBS.
3. Resuspend cells in 1 ml of complete culture medium. Proceed to cell counting with trypan blue exclusion to determine cell viability using a hemacytometer (*see* **Note 5**). Purify CD11c+ cells by positive selection by incubating with anti-CD11c magnetic beads according to the instructions of the selected manufacturer. Obtained purity should be higher than 90% (*see* **Note 6**).

4. Resuspend splenic CD11c⁺ cells at 4×10^5 cells/ml. Dispense 50 μl/well into a 96-well flat bottom microwell culture plates. Perform triplicates of the same condition. You can add 50 μl of a TLR ligand, such as LPS or Pam2CysK4 (previously diluted at 4 μg/ml in complete culture medium) (*see* **Note 7**). Otherwise, add 50 μl/well of complete culture medium. Incubate overnight at 37 °C and 5% CO_2.
5. Wash the cells with 200 μl of culture medium (*see* **Note 12**).
6. Resuspend the purified peritoneal $CD11c^+$ cells from infected animals (Subheading 3.1) in complete culture medium at 5×10^5 cells/ml. Dispense 100 μl/well into a 96-well flat bottom microwell culture plates.
7. Add 100 μl of purified $CD4^+$ T cells (1×10^6 cells/ml) from C57BL/6 mouse spleen previously purified with a mouse CD4 T cell Negative selection kit according to the manufacturer's instructions. The final volume is 200 μl/well. Incubate the cell culture for 5 days at 37 °C and 5% CO_2 (*see* **Note 13**).
8. Collect the culture supernatant and analyze the concentration of IFNγ, IL-4, IL-10 or IL-17 by specific sandwich ELISA (Fig. 3).

4 Notes

1. You can add 0.1% NaN_3 to the staining buffer for FACS in order to avoid bacterial and fungal growth.
2. To manipulate more comfortably and have more control over the animal, you can anesthetize the mouse with a low concentration of a ketamine–xylazine combination.
3. Cut the fur skin in the peritoneum without entering, in order to allow you to clearly see the injected liquid.
4. Make sure that ACK buffer is at room temperature to allow optimal cell lysis. When red blood cells have been lysed, the supernatant acquires a light red or pink color and the cell pellet turns white. 1 ml of ACK buffer and 1 min incubation period are conditions used for low to medium cell density (*e.g.* one spleen from a *naïve* mouse). In case that these conditions are not sufficient to completely lyse erythrocytes, repeat the incubation with an additional 1 ml ACK buffer.
5. Cell viability is calculated as the number of *viable cells* divided by the total number of cells within the grids on the hemacytometer. If cells take up trypan blue, they are considered nonviable.

6. The verification of the purity of the obtained CD11c$^+$ cells can be performed by flow cytometry, by staining with an anti-CD11c$^+$ antibody and evaluating the percentage of CD11c$^+$ cells.
7. Heat the LPS at 37 °C for 15 min for optimal stimulation of cells.
8. Culture supernatants can be stored at −20 °C up to 6 months.
9. If you analyze the cells by flow cytometry immediately after antibody staining, you can skip over the fixation step with 1% formaldehyde.
10. To analyze the expression of surface molecules on DCs you must first gate on CD11c$^+$ cells.
11. To improve isolation yield and purity of CD11c$^+$ from spleens, first incubate the spleen cut in small pieces in collagenase (1 mg/ml) and DNase I (20 μg/ml) in culture medium for 1 h at 37 °C.
12. To wash the cells, first centrifuge the plate at 320 × *g* for 2 min at 4 °C. Then, in sterile conditions, remove 180 μl of culture medium (that can be stored at −20 °C for analyses of cytokines) and add 200 μl of culture medium. Repeat this step twice.
13. Negative controls of your culture include (1) CD4$^+$ T cells incubated with CD11c$^+$ cells from infected mice, in absence of splenic CD11c$^+$ cells from *naïve* mice; (2) splenic CD11c$^+$ cells from *naïve* mice incubated with or without stimuli with CD11c$^+$ cells from infected mice in absence of CD4$^+$ T cells; and (3) coculture of CD11c$^+$ cells from *naïve* and CD4$^+$ T cells in medium (in the absence of stimuli).

References

1. Hewitson JP, Grainger JR, Maizels RM (2009) Helminth immunoregulation: the role of parasite secreted proteins in modulating host immunity. Mol Biochem Parasitol 167(1):1–11
2. Kane CM et al (2004) Helminth antigens modulate TLR-initiated dendritic cell activation. J Immunol 173(12):7454–7461
3. Rodriguez E et al (2017) Fasciola hepatica immune regulates CD11c(+) cells by interacting with the macrophage gal/GalNAc Lectin. Front Immunol 8:264
4. Rodriguez E et al (2017) Fasciola hepatica glycoconjugates immuneregulate dendritic cells through the dendritic cell-specific intercellular adhesion molecule-3-grabbing non-integrin inducing T cell anergy. Sci Rep 7:46748
5. Carasi P et al (2017) Heme-Oxygenase-1 expression contributes to the Immunoregulation induced by Fasciola hepatica and promotes infection. Front Immunol 8:883
6. Peron G et al (2018) Modulation of dendritic cell by pathogen antigens: where do we stand? Immunol Lett 196:91–102
7. Kapsenberg ML (2003) Dendritic-cell control of pathogen-driven T-cell polarization. Nat Rev Immunol 3(12):984–993
8. Dudek AM et al (2013) Immature, semi-mature, and fully mature dendritic cells: toward a DC-cancer cells Interface that augments anticancer immunity. Front Immunol 4:438
9. Lutz MB, Schuler G (2002) Immature, semi-mature and fully mature dendritic cells: which

signals induce tolerance or immunity? Trends Immunol 23(9):445–449
10. Dowling DJ et al (2010) Major secretory antigens of the helminth Fasciola hepatica activate a suppressive dendritic cell phenotype that attenuates Th17 cells but fails to activate Th2 immune responses. Infect Immun 78 (2):793–801
11. Falcon C et al (2010) Excretory-secretory products (ESP) from Fasciola hepatica induce tolerogenic properties in myeloid dendritic cells. Vet Immunol Immunopathol 137(1-2):36–46
12. Falcon CR et al (2012) Adoptive transfer of dendritic cells pulsed with Fasciola hepatica antigens and lipopolysaccharides confers protection against fasciolosis in mice. J Infect Dis 205(3):506–514
13. Hamilton CM et al (2009) The Fasciola hepatica tegumental antigen suppresses dendritic cell maturation and function. Infect Immun 77 (6):2488–2498

Chapter 14

Design of a Peptide-Carrier Vaccine Based on the Highly Immunogenic *Fasciola hepatica* Leucine Aminopeptidase

Cecilia Salazar, José F. Tort, and Carlos Carmona

Abstract

Many studies have shown that the degree of organization and repetitiveness of an antigen correlates with its efficiency to induce a B-cell response and production of neutralizing antibodies. Here we describe the design of a chimeric protein based on the hexamer form of the highly immunogenic *Fasciola hepatica* leucine aminopeptidase as a carrier system of small peptides with potential use as a multiepitope vaccine.

Key words Oligomer-peptide carrier system, Multiepitope vaccine, *Fasciola hepatica* juvenile

1 Introduction

The selection of molecules targeting key functions is a relevant aspect for controlling parasitic diseases. The availability of proteomes and/or extended genomics and transcriptomics information on parasitic species allows to search using bioinformatics tools for novel immunogens with special attention to the smaller noninfectious fragments of a pathogen with potential use in subunit vaccines. In addition, several studies have shown that the level of organization and repetitiveness of an antigen correlates with the efficiency to induced B-cell responses [1, 2]. Highly organized antigens induce B cells activation directly in the absence of T-cell help, as well as activating unresponsive (anergic) B cells. Moreover, nonorganized soluble antigen completely fails to induce a B-cell response [3]. Based on this, it has been proposed that large multimeric or repetitive antigens might act as carriers of small immunodominant peptides and resulting in stronger and more specific responses [4, 5]. The construction of chimeric proteins that might present diverse antigens is a novel avenue to be explored in the control of parasitic species.

In the case of the helminth parasite *Fasciola hepatica*, several antigenic proteins have been recognized experimentally. One of this

Martin Cancela and Gabriela Maggioli (eds.), *Fasciola hepatica: Methods and Protocols*, Methods in Molecular Biology, vol. 2137, https://doi.org/10.1007/978-1-0716-0475-5_14,

antigen is the leucine aminopeptidase (*Fh*LAP), a gut-associated digestive enzyme isolated from adult liver flukes, which has been proven to confer high levels of protection against fasciolosis in ruminants and this fact correlated with a strong humoral immune response [6, 7]. *Fh*LAP belongs to the M17 family of homohexameric peptidases that bind two metal cations. The enzyme is predicted to be cytoplasmic and folded in two domains. Like other members of the M17 family, *Fh*LAP C-terminal domain is conserved while the N-terminal portion of the protein is more divergent and similar to a restricted set of sequences that include other worms [8]. Due to its predicted structural features and ability to stimulate a specific immune response, *Fh*LAP has potential to be used as a chimeric immunogen, able to display repetitive predefined peptides on the molecular scaffold, such as other previously reported peptide-carrier systems [9, 10].

The success of a subunit vaccine highly depends on the immunogenicity of the target antigen. One of the challenges in developing these peptide-carrier systems is choosing the peptide/s to induce targeted immune responses. First and foremost, a key aspect is the identification of immunodominant epitopes that are capable of inducing protective immune response in terms of humoral immunity and/or cell mediated immunity against desired antigen [11]. Immunization with a specific epitope, corresponding to an effective neutralizing antibody, would be a "B-cell epitope-based vaccine" [12]. In addition, small peptides on its own are often weak immunogens and require not only carriers for delivery but adjuvants to enhance the immune response. To ensure specificity, the B-cell epitope chosen also will be unique to that protein to limit unwanted cross-reactivity of the generated antibodies with host-associated proteins [13].

Traditional B-cell epitope identification has depended upon experimental techniques, being costly and time-consuming [14]. In this regard, computational tools for epitope prediction have increased dramatically over the past years. Earlier computational methods based on single-scale amino acid propensity profiles tend to underperform on predicting linear B-cell epitopes [12]. Nowadays, linear B-cell epitope prediction methods combine two or more residue properties with machine learning approaches and more advanced algorithms (free available web servers are listed in Table 1). Despite this, the ability to predict epitopes from antigen sequences is still a complex task, and the accuracy of epitope prediction methods is still limited.

In the case of *F. hepatica* B-cell epitopes, most of the knowledge comes from studies involving epitope mapping methods such as the identification of linear B-cell epitopes recognized by polyclonal antibody from sheep vaccinated with glutathione S-transferase (*Fh*GST) [15]. Two dominant epitopes were identified in a saposin-like protein (*Fh*SAP2) using a set of overlapping

Table 1
Linear B-cell prediction and accessory tools

Tool	Search principle	URL	Protein input format	References
Sequence-based				
Propensity-scale tools (IEDB)		http://tools.iedb.org/bcell/	Plain text	
Chou and Fasman	Beta turn prediction			[36]
Emini	Surface accessibility			[35]
Karplus and Schulz	Flexibility scale			[34]
Kolaskar and Tongaonkar	Amino acid properties			[33]
Parker	Hydrophilicity			[37]
Machine learning-based				
ABCpred	Artificial neural network	http://crdd.osdd.net/raghava/abcpred/ABC.submission.html	Plain text	[45]
Bepipred 2.0	Random forest algorithm	http://www.cbs.dtu.dk/services/BepiPred/	FASTA format	[32]
LBTope*	Support vector machine	http://crdd.osdd.net/rafihava/lbtope/protein.php	FASTA format	[43]
BCPREDS**	Support vector machine	http://ailab.ist.psu.edu/bcpred/predict.html	Plain text	[38–41]
SVMTrip	Support vector machine	http://svsbio.unl.edu/SVMTriP/prediction.php	FASTA or plain text	[44]
COBEpro	Support vector machine	http://scratch.proteomics.ics.uci.edu/	Plain text	[42]
Alignment and visualization tools				
WebLogo		https://weblogo.berkeley.edu/	Protein Alignment	[64, 65]
Jalview		http://www.jalview.ore/	Several amino acid sequence formats/ Protein Alignment	[63]

* Two models of prediction: Fixed length of epitopes (model trained on 20mer epitopes): LBtope_Fixed, and LBtope_Fixed_non_redundant

Variable length of epitope (models trained on variable length of epitopes): LBtope_Variable, LBtope_Variable_non_redundant and LBtop_Confirm

** Two models of prediction: Fixed length epitope prediction:

BCPred and AAP Flexible length epitope prediction: FBCPred

synthetic peptides tested with sera from rabbits infected with this helminth [16]. Using a reverse vaccinology strategy, a range of *F. hepatica* B and T-cell epitopes were identified as immunogenic [17]. Also, synthetic peptide mimotopes based on the Cathepsin L1 (*Fh*CL1) protein sequence were identified, reducing fluke burden when used as vaccine antigens [18–21]. Linear B-cell epitopes using hydrophilicity, surface probability, flexibility, and antigenic index parameters were predicted for *Fh*CL1 [22], and recently, three regions of secreted *Fh*CL1 related to protection were identified using epitope mapping methods [23]. Also, a chimeric construction of a predicted linear B-cell epitope from *Fh*LAP and *Fh*CL1 was evaluated as a potential diagnostic tool in cattle [24].

*Fh*CL1 and *Fh*CL2 have already been recognized as promising vaccine candidates but these target antigens are mostly expressed in the adult stage of the liver fluke, when the liver damage has already taken place [25]. Controlling over the host's early immune response is fundamental for the parasite establishment, so targeting of juvenile stages through vaccination may prove beneficial. One of the juvenile-specific secreted proteases detected was the cathepsin L3 (*Fh*CL3), a cysteine protease with collagenolytic activity that has been suggested as major player in the migration in the host [26–28]. Therefore, targeting *Fh*CL3 function seems to be a reasonable strategy to prevent parasite establishment in the liver. Immunization with *Fh*CL3 elicited significant levels of protection [29]. Moreover, a region of the propeptide of the *Fh*CL3 suggested to be a potential target for protection in a rat infection model [30]. Here we describe an in silico approach to select a linear B-cell epitope of an antigen to be used as a peptide-carrier system with the highly immunogenic *Fh*LAP oligomer as a scaffold for a multiepitope vaccine against *F. hepatica*.

2 Materials

2.1 Data

For epitope prediction and homology modeling, the complete amino acid sequence of the protein of interest is selected. In our case we choose *Fh*CL3 (UniProt ID B3TM67) as a case of study. For peptide-carrier modeling, complete amino acid sequence of *Fh*LAP (UniProt ID Q17TZ3) and the primary sequence of the predicted *Fh*CL3 epitope are used.

2.2 Linear B-Cell Epitope Prediction Tools

It is generally believed that most of the identified linear antigenic determinants are part of conformational B-cell epitopes [31] and several online tools are available for the prediction of linear B-cell epitopes. Since this a complex and still unsettled task is always advisable to test different methods and compare results. Here we describe a few methods available that were used in our example case.

The Immune Epitope Database (IEDB, https://www.iedb.org) offers a useful interface for analysis and prediction of linear B cell epitopes in amino acid sequences, by collecting six methods and generating propensity scale profiles of the protein queries. The improved propensity scale BepiPred method predicts the location of linear B-cell epitopes using a combination of a Hidden Markov Model and propensity scale methods [32]. The Kolaskar–Tongaonkar antigenicity scale is used to predict antigenic determinants on proteins based on the physicochemical properties of amino acid residues. The frequency of occurrence in experimentally known segmental epitopes is also taken into consideration [33]. The Karplus method determines the flexibility scale based on mobility of protein segments [34]. Surface accessibility is also taken into consideration by Emini surface accessibility scale [35]. Experimental evidence has found that the beta turn and flexibility of proteins are also critical to B-cell epitope prediction, a feature exploited by the Chou and Fasman beta turn prediction tool [36] . Hydrophilicity is analyzed by the Parker hydrophilicity prediction method based in a hydrophilic scale generated on experimental data of peptide retention times during high-performance liquid chromatography (HPLC) on a reverse-phase column [37]. Although all these profiles cannot provide reliable predictions of linear B-cell epitopes, they could useful in highlighting potential antigenic regions of interest to confirm predictions made by other for example, BCPREDS [38].

BCPREDS is a machine learning-based tool that classifies amino acid peptide chains of specific lengths as either epitopes or nonepitopes. BCPREDS allows the user to select among three prediction methods: AAP method [39], BCPred [40], and FBCPred [41].

COBEpro also uses a Support Vector Machine (SVM) to make predictions on short peptide fragments within the query antigen sequence and then calculates an epitopic propensity score for each residue based on the fragment predictions [42]. LBtope [43] provides improved predictions of linear B-cell epitopes by training classifiers using experimentally validated nonepitopes, whereas all previous methods used randomly sampled fragments from UniProt as the nonepitope training data. SVMTriP [44] can improve the prediction performance for linear B-cell epitopes predicting epitopes with latest sequence input from IEDB database using support vector machines. SVM has been utilized by combining the Tripeptide similarity and Propensity scores in order to achieve the better prediction performance. A different approach using artificial neural networks is implemented by ABCpred [45]. This procedure predicts B-cell epitopes with 65.93% accuracy using a dataset for 700 B-cell epitopes and 700 non B-cell epitopes (random peptides) of maximum length of 20 residues. The B-cell epitope prediction tools are summarized in Table 1.

As mentioned initially the diverse methods rely on different approaches and the results most probably would be diverse. A useful assessment is to align all the results selecting the consensus regions highlighted by several different methods.

2.3 Homology Modeling Tools

2.3.1 I-TASSER Web Server

Using amino acid sequence as an input, I-TASSER [46–48] first generates three-dimensional (3D) atomic models from multiple threading alignments and iterative structural assembly simulations. The function of the protein is then inferred by structurally matching the 3D models with other known proteins. The output contains full-length secondary and tertiary structure predictions, and functional annotations on ligand-binding sites, Enzyme Commission numbers and Gene Ontology terms.

2.3.2 MODELLER Software

The input for MODELLER is an alignment of a sequence to be modeled with the template structure (s), the atomic coordinates of the template(s), and a simple script file. MODELLER then automatically calculates a model containing all nonhydrogen atoms, without any user intervention [49].

3 Methods

3.1 Overview

The focus of this section is obtaining the most reliable prediction of linear B-cell epitopes using diverse methods through free available web servers listed in Table 1. A relevant issue is to test diverse tools since predictions by different methods might differ. Consequently, a comparison of the results is mandatory, selecting the consensus region obtained by diverse predictors as a potential epitope (Fig. 1) For this purpose the protein alignment programs can be handy if they include instead of the amino acid sequence a profile of each prediction to be aligned. Once selected, a structural model of the epitope-carrier system can be generated using some of the homology modeling tools listed in Table 2.

3.2 B-Cell Epitope Prediction of the Target Protein

Given the amino acid sequence of the desired protein (e.g., *Fh*CL3), proceed to the diverse predictors to obtain a list of predicted linear B-cell epitopes within the target sequence:

1. Go to submission page of each of the B-cell prediction servers (Table 1). We recommend to check out all the features included in each web server, and paste the primary amino acid sequence of the target protein.
2. Select threshold and window length of the prediction or other parameters. Some tips for deciding on epitope length and other parameters are provided in **Note 1**.
3. Submit the sequence to obtain predicted B-cell epitopes in the target sequence.

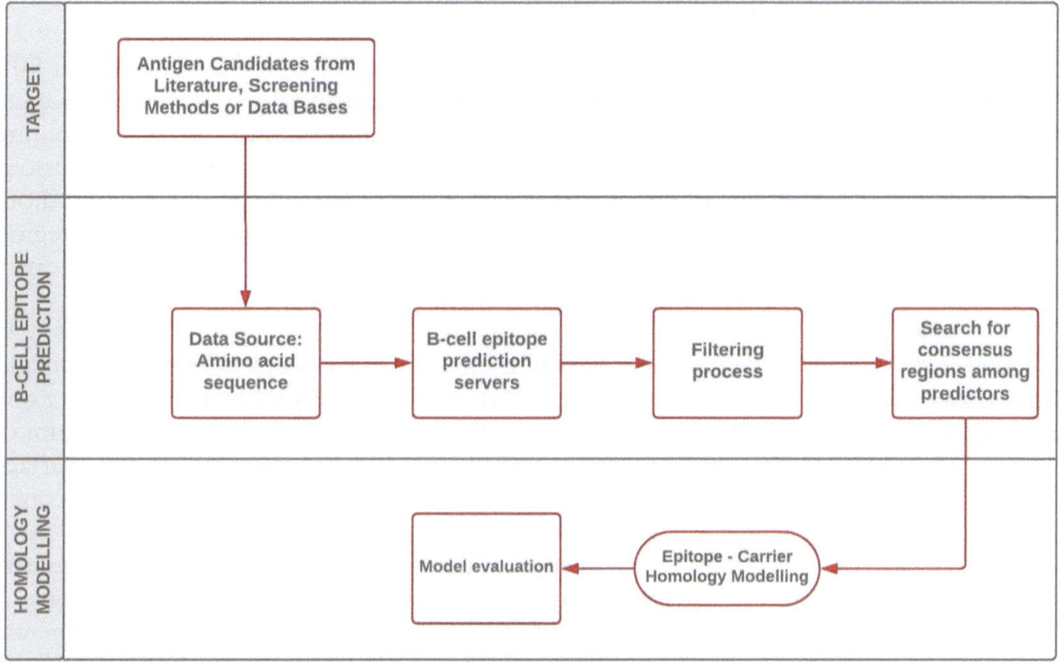

Fig. 1 Workflow for the theoretical design of the epitope-carrier antigen

Table 2
Homology modeling tools

Tool	URL	Input	References
Homology modeling			
I-TASSER web server	https://zhanglab.ccmb.med.umich.edu/I-TASSER/	Amino acid sequence	[46–48]
MODELLER software	https://salilab.org/modeller/	Target and template/s alignment, template/s PDB file	[49]
Model visualization and edit			
SPDB viewer software	http://www.expasy.org/spdbv/	PDB file	[51]
Model quality assessment			
SAVES v5	http://servicesn.mbi.ucla.edu/SAVES/	PDB file	[52–57]
Relative superficial accessibility			
NetSurfP	http: //www.cbs.dtu.dk/services/NetSurfP/	Amino acid sequence	[62]

4. Retrieve all the information of the predicted regions and perform a pairwise alignment or map the peptides against the target reference.

3.2.1 Evaluation of Predicted B-Cell Epitopes

The potential antigenic region should be predicted by at least 50% of the methods that recover verified epitopes and the region must not contain a conserved sequence also found in the host, avoiding possible antibody cross reactivity (Fig. 2). It is useful to filter out "false positives" such as epitopes predicted by a single method. Additionally, the predicted sequence must not contain a region comprised in the propeptide sequence, as it is usually cleaved from the native target protein (*see* **Note 2**).

3.3 Homology Modeling and Surface Accessibility

B-cell epitopes are often solvent expose portions of the antigen and in order to evaluate its exposure in the target protein structure one of the options is to map these regions against the 3D structure of the target protein and/or predict the overall Relative Surface Accessibility (RSA) score [50]. If the 3D structure of target protein is not listed in the RCSB Protein Data Bank (RCSB PDB, https://www.rcsb.org) an approximation of protein structure can be generate using comparative or homology modeling based in a single or multiple templates of resolved structures (*see* **Note 3**). To generate a target homology model based 3D structure using an online tool:

1. Go to submission page the homology modeling server (e.g., I-TASSER).
2. Paste the amino acid sequence of the target protein (e.g., *Fh*CL3) and select an option if necessary (*see* **Note 4**).
3. Submit the sequence to obtain a 3D model of the target protein (it usually takes a couple of hours to a couple of days).
4. Load the resulting PDB file on the Swiss PDB Viewer [51] and color the predicted regions on the 3D structure. Then, compute the surface of the target and color the surface exposure. A parallel RSA score can be computed from protein sequence using the NetSurfP tool [62].

3.3.1 Evaluation of Predicted Epitope Exposure

The predicted epitope should have high RSA score within the sequence and be located in an exposed surface of the 3D homology model.

3.4 Epitope-Carrier Assembly

1. To create a model for the chimeric construction, use the latest version of MODELLER and use a script to build a 3D model using information from multiple templates, in this case, the predicted structure of *Fh*CL3 and *Fh*LAP (*see* **Note 5**). Since *Fh*LAP structure is not listed in RCSB PDB, a similar approach

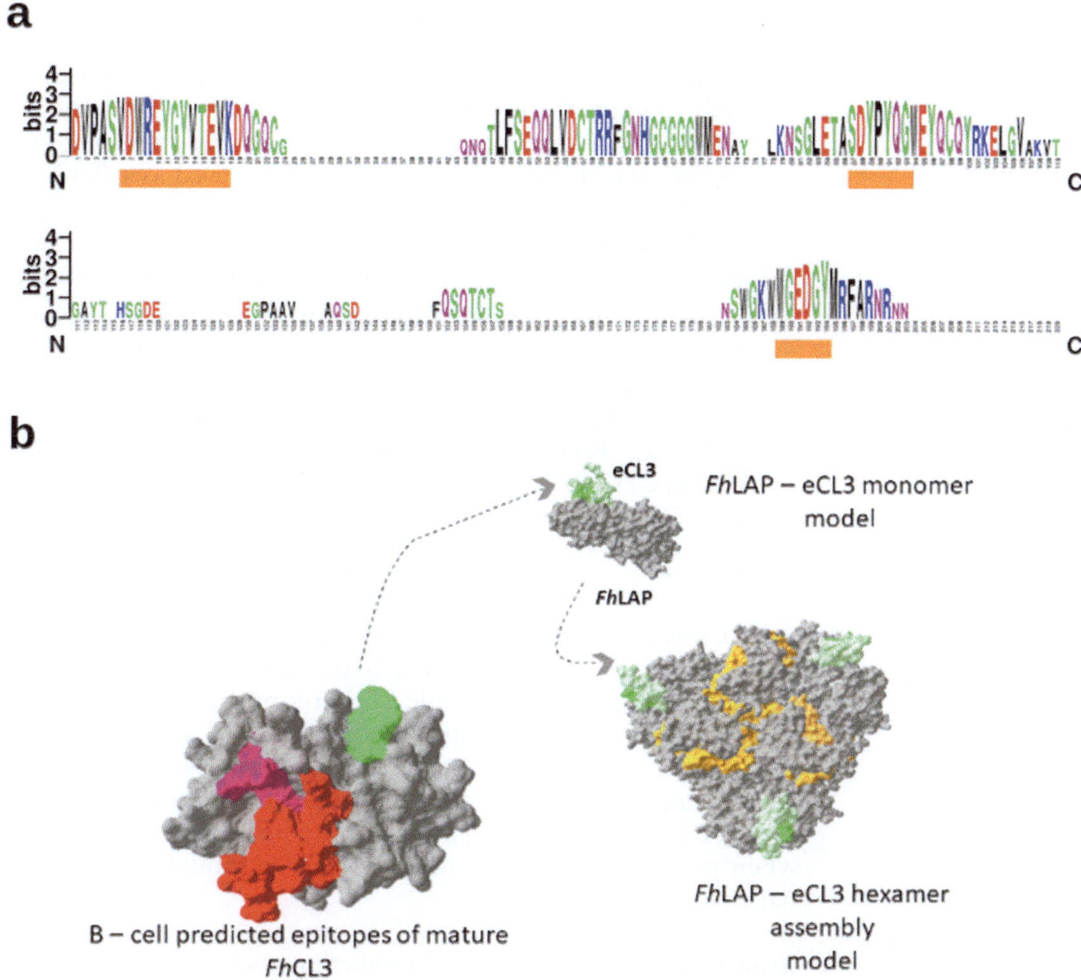

Fig. 2 Epitope selection and *Fh*LAP–eCL3 design. (**a**) Sequence logos of alignments and potential epitope regions from the mature *Fh*CL3 (marked in orange) using different online tools (Created with WebLogo). (**b**) Mapped potential epitopes on the 3D structure of the mature *Fh*CL3 and *Fh*LAP–eCL3 monomer and hexamer assembly models

for the generation of homology model of *Fh*CL3 can be performed for the monomer of *Fh*LAP.

2. Build the sequence structure of the epitope–carrier assembly (*see* **Note 6**).
3. For quaternary structure modeling, of the epitope–carrier hexamer, download and save a suitable leucine aminopeptidase oligomer template's coordinates as PDB file (e.g., bovine lens

leucine aminopeptidase, PDB ID 1LAM) and load it in the latest version of Swiss PDB Viewer [51]. Then perform an iterative magic fit with each subunit of the fusion protein on each subunit of the template chain in order to assemble the *Fh*LAP–eCL3 hexamer. The project containing the oligomer form of the chimeric construction can be saved as a PDB file. The Swiss PDB Viewer is applied for rendering resolved or predicted 3D structures. Also, you can edit visualization parameters and calculate surface exposure, among other features.

4. Evaluating the 3D models: reliability of the models generated can be assessed using several online tools. Saves v5: Structure Analysis and Verification server (http://servicesn.mbi.ucla.edu/SAVES/) runs in parallel six programs (Verify 3D [52, 53], ERRAT [54], Prove [55], PROCHECK [56], WHATCHECK [57], and CRYST (http://servicesn.mbi.ucla.edu/CRYST/) for checking and validating crystallographic structures and protein models before and after refinement (*see* **Note** 7).

In silico sequence-based homology models of the epitope-carrier protein is an important step to predict weather the expressed chimeric sequence has a suitable 3D conformation. Also, it allows to predict and evaluate the many possible spatial arrangements of different foreign peptide sequences insertions to the carrier protein and the impact of inserted sequences added by an expression vector during recombinant protein production. The key aspect to consider when generating a protein model is that is should resemble the native protein [58].

3.5 Final Considerations

Evaluation of performance of prediction tools is often difficult, especially when each of them has their own testing dataset. In order to assess if this approach is suitable for the chosen target protein, previous experimental verified epitopes within the target itself or closely related proteins should be recovered. In this case of the *Fh*CL3 is expected to recover two previously verified epitopes reported for CLs [30] as well additional predicted regions within the mature form of the protease (Fig. 2). Since the aim of these regions is to take part of a multiepitope vaccine formulation, we recommend to choose regions with the highest surface exposure in a nonstructured fragment of the protein sequence, in order to efficiently stimulate immune memory. Finally, in order to assess the immunogenicity of the predicted epitope and its potential use as a multiepitope vaccine using *Fh*LAP as a carrier, an immunization/challenge trial in a small experimental animal would determine its potential use as a vaccine.

4 Notes

1. Most of validated linear B-cell epitopes range in lengths between 12 and 16 amino acids so it seems reasonable to choose these lengths for predictions. It is recommend using low specificity cutoffs to avoid of missing some true positives epitopes and combining predictions from several tools to eliminate false positives, consensus predictions are usually more reliable than predictions obtained from a single prediction method [38].
2. To visualize the consensus regions among the prediction tools, one of the options is to create an alignment between the predicted peptide sequences and full length target, you can use Jalview 2.10.3b1 [63] or similar for computing and visualization. Also, if experimentally verified epitopes are available, you can use them as an internal control of the prediction methods. Additionally, it is worth to bear in mind that if a protein is found to be antigenic, it does not guarantee that it will be immunogenic [59]. Raising antibodies against a specific antigenic sequence could lead to neutralization of the antigen function, changes in biological activity as a consequence of a conformational change in the active site (or cofactor binding site) in the case of an enzyme, or it could also induce a modification in some portion of the protein, critical in determining the correct folding of the protein [60].
3. The 3D structure can be generated by any of the homology modeling web servers available (e.g., I-TASSER). See Homology Modeling section.
4. When using I-TASSER for homology modeling you can choose to add additional restrains and templates to guide the modeling process or to exclude templates from the I-TASSER library. When the sequence identity of the target structure is >40%, the homology models are satisfactory. As the percentage identity falls below 30%, model quality estimation on the basis of sequence identity becomes unreliable, as the relationship between sequence and structure similarity gets increasingly dispersed [61].
5. Visit https://salilab.org/modeller/manual/node21.html for an example on how to use multiple templates on MODELLER.
6. The comparison of the sequence obtained to other members of the M17 family showed that leucine aminopeptidases conserved domain is restricted the C-terminal portion while the N-terminal end of the protein is smaller and more variable. Since the C-terminal domain also bears the active site, the insertion of a foreign sequence might be less critical for the peptidase function and therefore for the assembly of the hexamer if the insert is added at the N-terminal end.

7. It is usually assumed that it is possible to synthesize effective vaccine immunogens by making short linear peptides adopt the structures observed in the native protein, but an isolated peptide sequence will not necessarily adopt the same conformation as in the native protein. Therefore, the antibodies raised against it may not recognize the same sequence in the native protein.

References

1. Bachmann MF, Rohrer UH, Kündig TM, Bürki K, Hengartner H, Zinkernagel RM (1993) The influence of antigen organization on B cell responsiveness. Science 262 (5138):1448–1451
2. Cooke MP, Heath AW, Shokat KM, Zeng Y, Finkelman FD, Linsley PS, Howard M, Goodnow CC (1994) Immunoglobulin signal transduction guides the specificity of B cell–T cell interactions and is blocked in tolerant self-reactive B cells. J Exp Med 179:425–438
3. Hinton HJ, Jegerlehner A, Bachmann MF (2008) Pattern recognition by B cells: the role of antigen repetitiveness versus toll-like receptors. Curr Top Microbiol Immunol 319:1–15
4. Sun JB, Holmgren J, Czerkinsky C (1994) Cholera toxin B subunit: an efficient transmucosal carrier-delivery system for induction of peripheral immunological tolerance. Proc Natl Acad Sci U S A 91(23):10795–10799
5. Laplagne DA, Zylberman V, Ainciart N, Steward MW, Sciutto E, Fossati CA, Goldbaum FA (2004) Engineering of a polymeric bacterial protein as a scaffold for the multiple display of peptides. Proteins 57(4):820–828
6. Piacenza L, Acosta D, Basmadjian I, Dalton JP, Carmona C (1999) Vaccination with cathepsin L proteinases and with leucine aminopeptidase induces high levels of protection against fascioliasis in sheep. Infect Immun 67(4):1954–61.
7. Maggioli G, Acosta D, Silveira F, Rossi S, Giacaman S, Basika T et al (2011) The recombinant gut-associated M17 leucine aminopeptidase in combination with different adjuvants confers a high level of protection against *Fasciola hepatica* infection in sheep. Vaccine 29 (48):9057–9063
8. Acosta D, Cancela M, Piacenza L, Roche L, Carmona C, Tort JF (2008) *Fasciola hepatica* leucine aminopeptidase, a promising candidate for vaccination against ruminant fasciolosis. Mol Biochem Parasitol 158(1):52–64
9. Rosas G, Fragoso G, Ainciart N, Esquivel-Guadarrama F, Santana A, Bobes RJ et al (2006) Brucella spp. lumazine synthase: a novel adjuvant and antigen delivery system to effectively induce oral immunity. Microbes Infect 8(5):1277–1286
10. Lee SF, Halperin SA, Salloum DF, MacMillan A, Morris A (2003) Mucosal immunization with a genetically engineered pertussis toxin S1 fragment-cholera toxin subunit B chimeric protein. Infect Immun 71 (4):2272–2275
11. Li W, Joshi M, Singhania S, Ramsey K, Murthy A (2014) Peptide vaccine: progress and challenges. Vaccine 2(3):515–536
12. Gershoni JM, Roitburd-Berman A, Siman-Tov DD, Freund NT, Weiss Y (2007) Epitope mapping: the first step in developing epitope-based vaccines. BioDrugs 21(3):145–156
13. Jones LH (2015) Recent advances in the molecular design of synthetic vaccines. Nat Chem 7(12):952–960
14. Blythe MJ, Flower DR (2005) Benchmarking B cell epitope prediction: underperformance of existing methods. Protein Sci 14(1):246–248
15. Sexton JL, Wilce MC, Colin T, Wijffels GL, Salvatore L, Feil S et al (1994) Vaccination of sheep against *Fasciola hepatica* with glutathione S-transferase. Identification and mapping of antibody epitopes on a three-dimensional model of the antigen. J Immunol 152 (4):1861–1872
16. Torres D, Espino AM (2006) Mapping of B-cell epitopes on a novel 11.5-kilodalton *Fasciola hepatica-Schistosoma mansoni* cross-reactive antigen belonging to a member of the F. hepatica saposin-like protein family. Infect Immun 74(8):4932–4938
17. Rojas-Caraballo J, López-Abán J, Pérez Del Villar L, Vizcaíno C, Vicente B, Fernández-Soto P et al (2014) In vitro and in vivo studies for assessing the immune response and protection-inducing ability conferred by *Fasciola hepatica*-derived synthetic peptides containing B- and T-cell epitopes. PLoS One 9(8): e105323
18. Villa-Mancera A, Reynoso-Palomar A, Utrera-Quintana F, Carreón-Luna L (2014) Cathepsin L1 mimotopes with adjuvant Quil a induces a Th1/Th2 immune response and confers significant protection against *Fasciola hepatica* infection in goats. Parasitol Res 113 (1):243–250

19. Villa-Mancera A, Quiroz-Romero H, Correa D, Alonso RA (2011) Proteolytic activity in *Fasciola hepatica* is reduced by the administration of cathepsin L mimotopes. J Helminthol 85(1):51–55
20. Villa-Mancera A, Méndez-Mendoza M (2012) Protection and antibody isotype responses against *Fasciola hepatica* with specific antibody to pIII-displayed peptide mimotopes of cathepsin L1 in sheep. Vet J 194(1):108–112
21. Villa-Mancera A, Quiroz-Romero H, Correa D, Ibarra F, Reyes-Pérez M, Reyes-Vivas H et al (2008) Induction of immunity in sheep to *Fasciola hepatica* with mimotopes of cathepsin L selected from a phage display library. Parasitology 135:1437–1445
22. Cornelissen JBWJ, Gaasenbeek CPH, Boersma W, Borgsteede FHM, Van Milligen FJ (1999) Use of a pre-selected epitope of cathepsin-L1in a highly specific peptide-based immunoassay for the diagnosis of *Fasciola hepatica* infections in cattle. Int J Parasitol 29 (5):685–696
23. Garza-cuartero L, Geurden T, Mahan SM, Hardham JM, Dalton JP, Mulcahy G (2018) Antibody recognition of cathepsin L1-derived peptides in *Fasciola hepatica* -infected and / or vaccinated cattle and identification of protective linear B-cell epitopes. Vaccine 36:958–968
24. Hernández-Guzmán K, Sahagún-Ruiz A, Vallecillo AJ, Cruz-Mendoza I, Quiroz-Romero H (2016) Construction and evaluation of a chimeric protein made from Fasciola hepatica leucine aminopeptidase and cathepsin L1. J Helminthol 90(1):7–13
25. Molina-Hernández V, Mulcahy G, Pérez J, Martínez-Moreno Á, Donnelly S, O'Neill SM et al (2015) *Fasciola hepatica* vaccine: we may not be there yet but we're on the right road. Vet Parasitol 208:101–111
26. Corvo I, O'Donoghue AJ, Pastro L, Pi-Denis N, Eroy-Reveles A, Roche L et al (2013) Dissecting the active site of the Collagenolytic Cathepsin L3 protease of the invasive stage of *Fasciola hepatica*. PLoS Negl Trop Dis 7(7):1–11
27. Corvo I, Cancela M, Cappetta M, Pi-Denis N, Tort JF, Roche L (2009) The major cathepsin L secreted by the invasive juvenile *Fasciola hepatica* prefers proline in the S2 subsite and can cleave collagen. Mol Biochem Parasitol 167(1):41–47
28. Robinson MW, Corvo I, Jones PM, George AM, Padula MP, To J et al (2011) Collagenolytic activities of the major secreted cathepsin L peptidases involved in the virulence of the helminth pathogen, *Fasciola hepatica*. PLoS Negl Trop Dis 5(4):e1012
29. Reszka N, Cornelissen JBWJ, Harmsen MM, Bieńkowska-Szewczyk K, De Bree J, Boersma WJ et al (2005) *Fasciola hepatica* procathepsin L3 protein expressed by a baculovirus recombinant can partly protect rats against fasciolosis. Vaccine 23(23):2987–2993
30. Harmsen MM, Cornelissen JBWJ, Buijs HECM, Boersma WJA, Jeurissen SHM, Van Milligen FJ (2004) Identification of a novel *Fasciola hepatica* cathepsin L protease containing protective epitopes within the propeptide. Int J Parasitol 34(6):675–682
31. Potocnakova L, Bhide M, Pulzova LB (2016) An introduction to B-cell epitope mapping and in Silico epitope prediction. J Immunol Res 2016:1–11
32. Jespersen MC, Peters B, Nielsen M, Marcatili P (2017) BepiPred-2.0: improving sequence-based B-cell epitope prediction using conformational epitopes. Nucleic Acids Res 45(W1): W24–W29
33. Kolaskar AS, Tongaonkar PC (1990) A semi-empirical method for prediction of antigenic determinants on protein antigens. FEBS Lett 276(1–2):172–174
34. Karplus PA, Schulz GE (1985) Prediction of chain flexibility in proteins. Naturwissenschaften 72(4):212–213
35. Emini EA, Hughes JV, Perlow DS, Boger J (1985) Induction of hepatitis a virus-neutralizing antibody by a virus-specific synthetic peptide. J Virol 55(3):836–839
36. Chou PY, Fasman GD (1978) Prediction of the secondary structure of proteins from their amino acid sequence. Adv Enzymol Relat Areas Mol Biol 47:45–148
37. Parker JMR, Guo D, Hodges RS (1986) New hydrophilicity scale derived from high-performance liquid chromatography peptide retention data: correlation of predicted surface residues with antigenicity and X-ray-derived accessible sites. Biochemistry 25 (19):5425–5432
38. El-Manzalawy Y, Dobbs D, Honavar VG (2017) In Silico prediction of linear B-cell epitopes on proteins. Methods Mol Biol 1484:255–264
39. Chen J, Liu H, Yang J, Chou K-C (2007) Prediction of linear B-cell epitopes using amino acid pair antigenicity scale. Amino Acids 33(3):423–428
40. El-Manzalawy Y, Dobbs D, Honavar V (2008) Predicting linear B-cell epitopes using string kernels. J Mol Recognit 21(4):243–255
41. El-Manzalawy Y, Dobbs D, Honavar V (2008) Predicting flexible length linear B-cell epitopes. Comput Syst Bioinformatics Conf 7:121–132

42. Sweredoski MJ, Baldi P (2009) COBEpro: a novel system for predicting continuous B-cell epitopes. Protein Eng Des Sel 22(3):113–120
43. Singh H, Ansari HR, Raghava GPS (2013) Improved method for linear B-cell epitope prediction using Antigen's primary sequence. PLoS One 8(5):1–8
44. Yao B, Zhang L, Liang S, Zhang C (2012) SVMTriP: a method to predict antigenic epitopes using support vector machine to integrate tri-peptide similarity and propensity. PLoS One 7(9):5–9
45. Saha S, Raghava GPS (2006) Prediction of continuous B-cell epitopes in an antigen using recurrent neural network. Proteins 65 (1):40–48
46. Yang J, Yan R, Roy A, Xu D, Poisson J, Zhang Y (2014) The I-TASSER suite: protein structure and function prediction. Nat Methods 12 (1):7–8
47. Roy A, Kucukural A, Zhang Y (2010) I-TASSER: a unified platform for automated protein structure and function prediction. Nat Protoc 5(4):725–738
48. Zhang Y (2008) I-TASSER server for protein 3D structure prediction. BMC Bioinformatics 9:1–8
49. Sali A, Blundell TL (1993) Comparative protein modelling by satisfaction of spatial restraints. J Mol Biol 234(3):779–815
50. Petersen B, Petersen TN, Andersen P, Nielsen M, Lundegaard C (2009) A generic method for assignment of reliability scores applied to solvent accessibility predictions. BMC Struct Biol 9:1–10
51. Guex N, Peitsch MC (1997) SWISS-MODEL and the Swiss-PdbViewer: an environment for comparative protein modeling. Electrophoresis 18(15):2714–2723
52. Bowie JU, Luthy R, Eisenberg D (1991) A method to identify protein sequences that fold into a known three-dimensional structure. Science 253(5016):164–170
53. Luthy R, Bowie JU, Eisenberg D (1992) Assessment of protein models with three-dimensional profiles. Nature 356 (6364):83–85
54. Colovos C, Yeates T (1993) Verification of protein structures: patterns of nonbonded atomic interactions. Protein Sci 2:1511–1519
55. Pontius J, Richelle J, Wodak SJ (1996) Deviations from standard atomic volumes as a quality measure for protein crystal structures. J Mol Biol 264(1):121–136
56. Laskowski RA, MacArthur MW, Moss DS, Thornton JM (1993) PROCHECK: a program to check the stereochemical quality of protein structures. J Appl Crystallogr 26(2):283–291
57. Hooft RW, Vriend G, Sander C, Abola EE (1996) Errors in protein structures. Nature 381, 272
58. Lam SD, Das S, Sillitoe I, Orengo C (2017) An overview of comparative modelling and resources dedicated to large-scale modelling of genome sequences. Acta Crystallogr D Struct Biol 73(8):628–640
59. Van Regenmortel MHV (2001) Antigenicity and immunogenicity of synthetic peptides. Biologicals 29(3–4):209–213
60. Celada F, Strom R (1972) Antibody-induced conformational changes in proteins. Q Rev Biophys 3:395–425
61. Fiser A (2010) Template-based protein structure modeling. Methods Mol Biol 673:73–94
62. Michael SK, Martin CJ, Henrik N, Kamilla KJ, Vanessa IJ, Casper KS, Morten OAS, Ole W, Morten N, Bent P, Paolo M (2019) NetSurfP-2.0: Improved prediction of protein structural features by integrated deep learning. Proteins: Structure, Function, and Bioinformatics 87 (6):520–527
63. Waterhouse AM, Procter JB,Martin DMA, Clamp M, Barton GJ (2009) Jalview Version 2--a multiple sequence alignment editor and analysis workbench. Bioinformatics 25 (9):1189–1191
64. G. E. Crooks, Hon G, Chandonia JM, Brenner SE (2004) WebLogo: A Sequence Logo Generator. Genome Research 14 (6):1188–1190
65. Schneider TD, Stephens RM (1990) Sequence logos: a new way to display consensus sequences. Nucleic Acids Research 18 (20):6097–6100

Chapter 15

Liver Fluke Vaccine Assessment in Cattle

Gabriela Maggioli, Cecilia Salazar, Federico Fossa, and Carlos Carmona

Abstract

Liver fluke *Fasciola hepatica* remains an important agent of foodborne trematode disease producing great economic losses due to its negative effect on productivity of grazing livestock in temperate areas. The prevailing control strategies based on antihelminthic drugs are not long term sustainable due to widespread resistance. Hence, vaccination appears as an attractive option to pursue for parasite eradication.

Key words *Fasciola hepatica*, Vaccine, Livestock

1 Introduction

F. hepatica, a trematode parasite with a worldwide distribution, causes massive economic losses and animal health problems in livestock. Fasciolosis is a zoonotic infection and has been identified as a reemerging neglected tropical disease by WHO [1]. Moreover, there is and estimation that 180 million of people are at risk of infection mostly in South America and Africa [2].

Economic losses in agriculture are related to reduction in meat, milk, and wool output, costs associated with antihelminthic drug control and secondary bacterial infections. These losses were previously estimated at US$ 3 billion annually [3], but are likely to be far higher currently [4–8].

Triclabendazole (TCBZ) is the drug of choice for the control of acute and chronic fasciolosis in ruminants [9–11]. However, the overreliance on TCBZ to treat ruminants especially sheep and, to a lesser extent, cattle, has resulted in selection of TCBZ-resistant flukes [12]. Moreover, antihelminthic residues in meat and milk have great impediments to the use of these products for human consumption [13, 14]. In this context, vaccination emerges as an alternative capable of providing both long-term protection and sustainability. Also, a vaccine approach circumvents not only the human consumption of antihelminthic residues problem, but also emerges as an environment friendly option [15]. Indeed, vaccine

Martin Cancela and Gabriela Maggioli (eds.), *Fasciola hepatica: Methods and Protocols*, Methods in Molecular Biology, vol. 2137, https://doi.org/10.1007/978-1-0716-0475-5_15,

trials with significant levels of protection in ruminants using a combination of antigens, suggest a fluke vaccine is a feasible method for the control of fasciolosis in livestock that will lessen antihelminthic use and slow the spread resistance [16].

Ruminants and humans are infected by ingestion of the metacercariae encysted in vegetation. Inside the duodenum the juvenile (NEJ) emerge and rapidly penetrate throughout the gut wall, then the migratory stage of the fluke migrates through the liver and establishes in the bile ducts where it reaches sexual maturity 8–10 weeks postinfection. The hermaphrodite worms release embryonated eggs in the feces [17]. The results obtained in pilot studies using murine models have been difficult to replicate in livestock species. Hence, the use of natural hosts facilitates the assessment of vaccine performance and the underlying mechanisms by which host immune response occur during the early and late stages of the helminth parasite and allows to assess the immune response and protection parameters developed by vaccination. In this chapter, we describe a vaccination protocol against *F. hepatica* in a ruminant infection model.

2 Materials

2.1 Vaccination Procedure

2.1.1 Animal Model

1. 10–12 weaner calves per group from a farm with no liver fluke infection history (*see* **Notes 1** and **2**).
2. Cattle (below 12 months old) are divided into control and experimental groups.
3. Control groups are subdivided into control 1, cattle with Immunization with phosphate-buffered saline (PBS) and infection; control 2, cattle with immunization with adjuvant and infection.
4. Livestock Scales.

2.1.2 Antigen and Vaccine Formulation

1. Antigen: 1 mg of recombinant *F. hepatica* Leucine aminopeptidase (*Fh*LAP), filtered by 0.22 μm.
2. 0.01 M phosphate-buffered saline (PBS) sterilize by autoclaving.
3. Adjuvant: Adyuvac 50 (Virbac-Santa Elena, Uruguay).
4. 5 mL Syringe.
5. Hypodermic Needles 20GX11/2″; 0.9 × 38 mm.

2.1.3 Serum Samples

1. 20 mL syringe.
2. 15 mL conical tubes.
3. Centrifuge up to 4000 × *g*.
4. Incubator 37 °C.

2.1.4 Oral Challenge

1. Cellulose filter paper.
2. Fresh metacercariae.
3. Steel feeding tube.

2.2 Worm Recovery

1. Knife.
2. Trays.
3. PBS at 37 °C.
4. Pail.
5. Test Sieve 600 μm.
6. Petri dish.
7. Stereomicroscope.

2.3 Peripheral Blood Mononuclear Cell (PBMC) Isolation

1. 20 mL syringe.
2. 15 mL conical tubes containing Tris–HCl 50 mM, pH 8, EDTA 1%.
3. Centrifuge reaching up to 4000 × *g*.
4. Biosafety cabinet.
5. 15 mL sterile conical tubes.
6. 15 mL sterile conic tubes containing 4 mL of commercial lymphocyte separation medium (1.077 g/mL).
7. Sterile PBS.
8. Sterile red blood cell lysis buffer (RBC): 155 mM NH_4Cl, 10 mM $KHCO_3$, 0.1 mM EDTA.
9. RNA isolation reagent.

2.4 Detection of Antibody Levels

1. 96-well microtiter plate.
2. PBS sterilized by autoclaving.
3. PBS-T: PBS, 0.05% Tween 20.
4. Block Solution: PBS, 0.1% Tween 20, 5% skimmed milk.
5. Dilution Solution: PBS-T, 2.5% skimmed milk.
6. HRP-conjugated rabbit anti-bovine IgG (whole molecule), HRP-conjugated mouse anti-bovine IgG1 or IgG2.
7. 0.05 M phosphate-citrate buffer: To 900 mL of distilled H_2O, add 362 mg Na_2HPO_4 and 126 mg citric acid. Adjust pH to 5.0. Adjust to 1 L with distilled H_2O.
8. 30% (v/v) H_2O_2 solution.
9. *o*-phenylenediamine (OPD) substrate solution: Dissolve OPD (0.04%) in 0.05 M phosphate-citrate buffer. Add 6 μL of fresh 30% H_2O_2 per 10 mL of OPD substrate solution immediately prior to use.
10. 1 M HCl.
11. An automatic spectrophotometer: filter $\lambda = 492$ nm.

3 Methods

3.1 Vaccination Procedure

1. Vaccination protocol is shown in Fig. 1 (*see* **Note 1**).
2. Before the beginning of trial, all animals need to be tested to sort out any possible previous encounter with *F. hepatica* by coproantigen and serology methods (*see* **Note 2**).
3. Prior receiving the immunization schedule, weight each animal.
4. Collect approximately 5 mL of whole blood and 5 mL of blood in an EDTA tube by jugular venipuncture along the trial (*see* **Note 3**).
5. Prepare the vaccine formulation by mixing 1 mg of FhLAP and the appropriate amount of Adyuvac 50 adjuvant (*see* **Note 4**).
6. Inject the emulsion (up to 2 mL) subcutaneously into the neck area behind the ears.
7. The animal's overall health should be inspected by a veterinarian 3 days after each vaccination and regularly along the trial.
8. Four weeks after the last vaccine dose, cattle are orally challenged with 400–600 fresh metacercariae contained in a 0.5 cm handmade cellulose paper cylinder and delivered using a feeding tube (*see* **Note 5**).
9. Animals are kept in suitable condition for up to 10–12 weeks after the challenge until the end point of the trial.

3.2 Worm Recovery

1. At 10–12 weeks after infection, animals are euthanized in a local abattoir and livers are collected and properly identified (*see* **Note 6**).
2. Place the liver in a tray and register images using a digital camera (*see* **Note 7**).
3. Remove the gallbladder attached to the livers to obtain *F. hepatica* eggs to evaluate effect on egg output and egg viability (*see* Chapter 1, Subheadings 3.2 and 3.3).
4. Cut the liver in small sections (approximately 1 cm width). Recover any visible worm. Then, incubate the pieces of liver in a pail with PBS at 37 °C during 1 h (*see* **Note 8**).

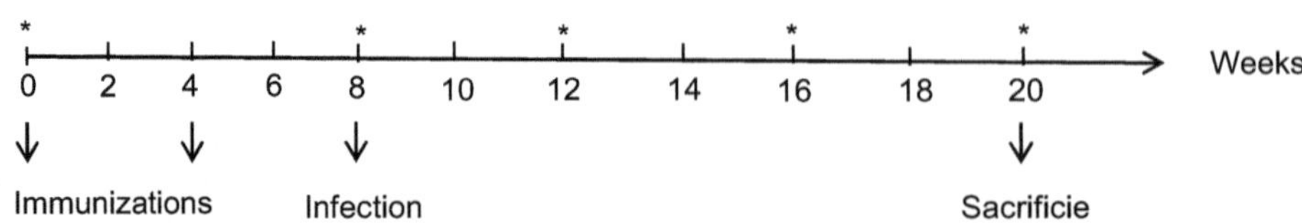

Fig. 1 Scheme of vaccination protocol. Cattle are immunized two times (week 0 and 4) and infected with 400–600 metacercariae per cattle at week 8. (*) Control cattle weight. Blood is collected at 2 week intervals during the study

5. Remove the pieces of liver and pass the PBS through the test sieve. Examine under stereomicroscope the material retained in the sieve.
6. Calculate the percentage of protection using the formula below:

   %Protection = [(C – I)/C] × 100

   C: worm recovery from the control group.

   I: worm recovery from immunized group.

3.3 PBMC Isolation

1. Collect 5 mL of bovine blood sample in a 15 mL conical tube containing 1% EDTA and mix well by inverting the tube.
2. Dilute the blood sample in sterile PBS (1:2) in a biosafety cabinet.
3. Gently layer the diluted blood on the top the lymphocyte separation medium using a 1 mL pipette. Note that the blood and separation medium should stay as two different layers.
4. Centrifuge the tubes for 40 min at 400 × *g* in a 4 °C swing-out bucket with no breaks.
5. Collect the PBMCs by aspiration of the cloudy coat (approximately 1 mL) corresponding to the interphase between the separation medium and the soluble phase. Add 10 mL of PBS and centrifuge for 15 min at 100 × *g* (standard acceleration/deceleration settings).
6. Incubate the pellet with 1 mL of RBC lysis buffer at room temperature for 3 min. Stop the reaction using 10 mL of PBS and centrifuge at 100 × *g*. Wash the pellet with sterile PBS and centrifuge at 100 × *g* 15 min (*see* **Note 9**).
7. For RNA isolation, add the recommended amount of the RNA isolation reagent and proceed to RNA extraction or store at −80 °C until use (*see* **Note 10**).

3.4 Detection of Antibody Levels

1. After blood collection, allowed to clot for 30–60 min at 37 °C. Place the clot at 4 °C overnight to allow it to contract. The serum should then be removed from the clot by centrifugation at 4000 × *g* for 10 min at 4 °C. Serum can be stored for many years at −20 °C or below.
2. Coat the 96-well microtiter plate with 100 μL of 2 μg/μL *Fh*LAP diluted in PBS and incubate at 37 °C for 1 h (*see* **Note 11**).
3. Wash the wells with PBS-T, three times for 5 min each.
4. Add 200 μL of blocking solution per each well and incubate overnight at 4 °C (*see* **Note 12**).
5. Wash the wells with PBS-T three times for 5 min each.

6. Add 100 μL of the sera diluted with dilution solution to each well and incubate at 37 °C for 1 h (*see* **Note 12**).
7. Wash with PBS-T three times for 5 min each.
8. Add 100 μL per well of HRP-conjugate anti-bovine IgG whole molecule or IgGl or IgG2 diluted in 'dilution solution' (appropiate dilution). Then incubate at 37 °C for 1 h (*see* **Note 13**)
9. Wash the wells with PBS-T three times for 5 min each.
10. Add 100 μL of OPD substrate solution to each well and incubate at room temperature for 10 min in the dark.
11. Add 50 μL of 1 M HCl into the well to stop the enzymatic reaction.
12. Measure the optical density at 492 nm (OD_{492}) in an automatic spectrophotometer (*see* **Note 14**).

4 Notes

1. Animals should be acquired from farms with no history of *F. hepatica* infection. If possible, test all animals by a fecal egg count assay for any possible infection with *F. hepatica* and other helminth parasite. An antihelminthic treatment might be necessary in case of detection of any gastrointestinal nematode or cestode. In case the test comes positive for *F. hepatica* infection, exclude the animal from the experiment. The vaccination scheme should be applied to wiener beef calves weighed 120–140 kg. Ideally, animals should be vaccinated before grazing. Also, a ≥ 21-day acclimatization period is recommended.
2. Collect individual fecal samples for helminth egg counts (FEC) and liver fluke coproantigen using an ELISA commercial kit in order to confirm infection status. Collect serum from blood for fluke antibody detection by ELISA. The are many available methods to test *F. hepatica* infection, a recent publication compared some of these methods [18].
3. We recommend biweekly blood sampling for specific antibody detection (total IgG, IgG1 and IgG2 isotypes). A longitudinal analysis provides comparative data over time and vaccine performance overview.
4. Mix the proteins with selected adjuvant according to the manufacturer's instructions. Many adjuvants are available to use for the recombinant protein vaccination. Selecting an appropriate adjuvant for each protein and animal models is a topic of considerable relevance.

5. Monitor specific antibody levels production prior to infection. Use min 400 and max 600 fresh metacercariae. This number can be variable depending on metacercariae viability, thus is highly recommended to determine the percentage of viable metacercariae by in vitro excystment assay (>70% is recommended) (*see* Chapter 1, Subheading 3.6). For each infection dose, collect the necessary amount of metacercariae in a 200 μL H_2O volume and into a cellulose paper cylinder. Allow the filter paper to set overnight at 4 °C. Use a bovine stainless steel feeding tube to deliver the dose.
6. End point is determined based on the assumption that most of the worms have reached maturity and are established in the liver.
7. All images are analyzed and lesions are scored as followed: 0: absent; 1: mild; 2: moderate; 3: severe; 4: very severe.
8. Incubate into warm PBS to recover most of parasites. We let the worms migrate out from the dissected liver.
9. It is highly recommended to inspect the PBMC fraction under the microscope to rule out any contamination. Also, it is recommended to perform a cell count and viability check using standard methods.
10. High quality RNA extraction assessed using for example Bioanalyzer equipment is highly recommended previous library preparation for transcriptional profile analysis (such as RNA-seq) or cDNA synthesis for quantitative PCR (qPCR) of selected target genes.
11. Dilution of the antigen and the HRP-conjugated sheep anti-bovine IgG, IgG1 or IgG2 used in ELISA can be determined by varying dilutions prior to obtain the optimal values. The suggested HRP-conjugate antibodies dilution provided by the manufacturer's instructions is an optimal reference.
12. To block the nonspecific binding in ELISA.
13. Serum dilutions (single dilution or threefold starting at 1/10) were added in duplicate to the plate.
14. For sera threefold dilutions the antibody titers were calculated as the reciprocal of that dilution of serum falling midway on the linear portion of the OD curve of a range of positive control sera.

References

1. WHO (2018) Neglected tropical diseases. Human fascioliasis: review provides fresh perspectives on infection and control. who.int/foodborne_trematode_infections/fascioliasis/en/
2. Carmona C, Tort JF (2017) Fasciolosis in South America: epidemiology and control challenges. J Helminthol 91(2):99–109. https://doi.org/10.1017/S0022149X16000560

3. Spithill TW, Smooker PM, Copeman DB (1999) Fasciola gigantica: epidemiology, control, immunology and molecular biology. In: Dalton JP (ed) Fasciolosis. CAB International Publishing, Wallingford, Oxon, pp 377–410
4. Schweizer G, Braun U, Deplazes P, Torgerson PR (2005) Estimating the financial losses due to bovine fasciolosis in Switzerland. Vet Rec 157:188–193
5. Kaplan RM (2001) Fasciola hepatica: a review of the economic impact in cattle and considerations for control. Vet Ther 2(1):40–50
6. Charlier J, Vercruysse J, Morgan E, van Dijk J, Williams DJ (2014) Recent advances in the diagnosis, impact on production and prediction of Fasciola hepatica in cattle. Parasitology 141(3):326–335. https://doi.org/10.1017/S0031182013001662
7. Andrews S (1999) In: Dalton JP (ed) The life cycle of Faciola hepatica. En: Fasciolosis. CAB International, Wallingford, pp 1–20
8. Hatschbach PJ (1995). A Fasciola hepatica a sua historia. A Hora Vet (Ed.Extra) 15 (1): 10–11
9. Fairweather I, Boray JC (1999) Fasciolicides: efficacy, actions, resistance and its management. Vet J 158(2):81–112
10. Brennan GP, Fairweather I, Trudgett A, Hoey E et al (2007) Understanding triclabendazole resistance. Exp Mol Pathol 82(2):104–109
11. Fairweather I (2011) Liverfluke isolates: a question of provenance. Vet Parasitol 176:1–8
12. Kelley JM, Elliott TP, Beddoe T, Anderson G, Skuce P, Spithill TW (2016) Current threat of triclabendazole resistance in Fasciola hepatica. Trends Parasitol 32:458–469
13. Toet H, Piedrafita DM, Spithill TW (2014) Liver fluke vaccines in ruminants: strategies, progress and future opportunities. Int J Parasitol 44:915–927
14. Imperiale F, Ortiz P, Cabrera M, Farias C, Sallovitz JM, Iezzi S, Pérez J, Alvarez L, Lanusse C (2011) Residual concentrations of the flukicidal compound triclabendazole in dairy cows' milk and cheese. Food Addit Contam Part A Chem Anal Control Expo Risk Assess 28 (4):438–445. https://doi.org/10.1080/19440049.2010.551422
15. Molina-Hernández V, Mulcahy G, Pérez J, Martínez-Moreno A, Donnelly S, O'Neill S, Daltona JP, Cwiklinski K (2015) Fasciola hepatica vaccine: we may not be there yet but we're on the right road. Vet Parasitol 208:101–111
16. Beesley NJ, Caminade C, Charlier J, Flynn RJ, Hodgkinson JE, Martinez-Moreno A, Martinez-Valladares M, Perez J, Rinaldi L, Williams DJL (2018) Fasciola and fasciolosis in ruminants in Europe: identifying research needs. Transbound Emerg Dis 65(Suppl 1):199–216. https://doi.org/10.1111/tbed.12682
17. Dalton JP (1999) Fasciolosis. CABI Pub, Wallingford
18. Calvani NED, George SD, Windsor PA, Bush RD, Šlapeta J (2018) Comparison of early detection of Fasciola hepatica in experimentally infected merino sheep by real-time PCR, coproantigen ELISA and sedimentation. Vet Parasitol 251:85–89. https://doi.org/10.1016/j.vetpar.2018.01.004

Chapter 16

Testing Albendazole Resistance in *Fasciola hepatica*

Luis I. Alvarez, María Martinez Valladares, Candela Canton, Carlos E. Lanusse, and Laura Ceballos

Abstract

The egg development test is a useful in vitro tool to detect albendazole (ABZ) resistance in *Fasciola hepatica*. ABZ is the only flukicidal compound with ovicidal activity. The described test is based on the ABZ capacity to affect parasite egg development and hatching in susceptible parasites, while this effect is lost in ABZ-resistant liver fluke isolates. Among many advantages, it is noted that the diagnostic test can be performed on eggs isolated from fecal samples (sheep and cattle), avoiding the sacrifice of animals necessary in controlled efficacy trials. The egg development test described here is a simple, inexpensive, and accessible method, previously employed for diagnosis of ABZ resistance in *F. hepatica*.

Key words Albendazole, *Fasciola hepatica*, Egg development test, Ovicidal activity, Resistance, Diagnostic

1 Introduction

Chemotherapy based on the use of flukicidal compounds is the main tool to control the liver trematode *Fasciola hepatica*, affecting livestock animals and humans. Benzimidazoles (BZD) are broad-spectrum anthelmintic compounds widely used in human and veterinary medicine to control nematode, cestode, and trematode infections [1]. Although triclabendazole is the main flukicidal used drug, albendazole (ABZ) is the only BZD methylcarbamate recommended to control fascioliasis in domestic animals, although its activity is restricted to flukes older than 12 weeks [1]. The frequent use of effective flukicidal compounds has resulted in an increased selection for ABZ-resistant flukes [2–4]. The standard and well-established protocol for the determination of drug activity against *Fasciola* spp. in ruminants is the controlled efficacy test [5]. In this test, the efficacy is determined by comparing the number of flukes recovered after the slaughter of infected animals treated with a drug or placebo. However, this assay is highly expensive and time-consuming. The fecal egg count reduction test is an

Martin Cancela and Gabriela Maggioli (eds.), *Fasciola hepatica: Methods and Protocols*, Methods in Molecular Biology, vol. 2137, https://doi.org/10.1007/978-1-0716-0475-5_16,

alternative method to determine the efficacy of a treatment, or the drug susceptibility of an *F. hepatica* isolate. The efficacy is declared if a reduction ≥95% of fecal fluke egg counts at 14 days post-treatment is achieved [5]. However, the release of eggs stored in the gall bladder after an effective treatment may lead to false-positive results [6]. This fact emphasizes the need of more robust, sensitive and accurate method to determine the resistance status of *F. hepatica* populations and achieve an optimal use of anthelmintics in farms.

The antiparasitic activity of BZD methylcarbamates largely depends on their affinity for parasite ß-tubulin, leading to subsequent disruption of the tubulin–microtubule dynamic equilibrium [7]. In nematodes and trematodes, this particular mode of action explains the effect of BZD on egg development and hatching [8, 9]. Based on this BZD capacity, the egg development test was developed for the diagnosis of ABZ resistance in *F. hepatica* [10, 11]; eggs from resistant isolates develop and hatch at a higher drug concentration than those from susceptible isolates. The protocol described in the current chapter is a simple, inexpensive and accessible method. Therefore, the egg development test can be considered as a helpful and original tool for the diagnosis of ABZ resistance in *F. hepatica* populations from both domestic animals and humans.

2 Materials

1. *F. hepatica* eggs: eggs obtained from infected animals (i.e., sheep, cattle) recovered either directly from the gallbladder (after slaughter) or from fecal material (*see* **Note 1**).
2. Albendazole (ABZ) pure standard (*see* **Note 2**).
3. Methanol (MEOH) or dimethyl sulfoxide (DMSO) of analytical grade (to dissolve pure drug).
4. Incubator.
5. Kahn culture tubes (5 mL).
6. Microscope (*see* **Note 3**).
7. Pasteur pipettes (1.5 mL).
8. Tap water (for egg washing).
9. Aluminum foil.
10. Cover slides.
11. Conical centrifugal tubes (15 and 50 mL).
12. Precision pipettes (to dispense 1 mL and 10 μL volume).
13. 10% (v/v) buffered formalin.
14. Precision analytical balance (to weight ABZ).

15. 5000 μM (5000 nmoles/mL) ABZ solution: weigh 13.26 mg of pure ABZ and dissolve it in 10 mL of MEOH. DMSO can also be used as drug solvent to replace MEOH. Store at −20 °C until use (*see* **Note 4**).
16. ABZ working solutions: dilution of the ABZ solution in MEOH (or DMSO) to achieve final concentrations of 500 μM (500 nmoles/mL), 50 μM (50 nmoles/mL), and 5 μM (5 nmoles/mL). Store at −20 °C until use (*see* **Note 5**).

3 Methods

3.1 Egg Incubation

1. Prepare a *F. hepatica* eggs suspension (approx. 10 mL) of 200 eggs/mL in tap water (*see* **Note 6**).
2. After homogenization, place 1 mL of the egg suspension in Kahn culture tubes (200 eggs/mL).
3. The egg development test is performed including five replicates for each ABZ concentration to be tested. The egg development is assessed at the concentration of 5, 0.5, and 0.05 μM, which are reached by addition of 10 μL of each ABZ working solution (500, 50, and 5 μM, respectively). A set of untreated eggs ($n = 5$) needs to be included as control. Add 10 μL of the solvent used for ABZ dissolution in the working solution (MEOH or DMSO) to the untreated eggs (*see* **Note** 7).
4. Incubate eggs at 25 °C in darkness for a 12 h period (*see* **Note 8**).
5. After the drug incubation period, wash either untreated or treated eggs with tap water to facilitate drug/solvent removal ("eggs washing step") (*see* **Note 9**).
6. Keep washed eggs in 1 mL of tap water in darkness at 25 °C over 15 days.

3.2 Egg Development Test

1. After the 15 day incubation period, expose the eggs to daylight for 4 h. Eggs can also be exposed 4 h to light from an artificial source of at least 1000 lumens (lm).
2. Immediately afterwards, add 10 μL of 10% (v/v) buffered formalin to each tube and mix properly to stop egg development.
3. Pipet carefully approximately 0.2 mL from the bottom of each tube to transfer sediment eggs in a cover slide in order to proceed to microscopic evaluation.
4. Evaluate developed and nondeveloped eggs by visual inspection through optical microscope (40× magnification). Count approximately 100–150 eggs to estimate the proportion of developed and nondeveloped eggs in each tube (*see* **Note 10**).

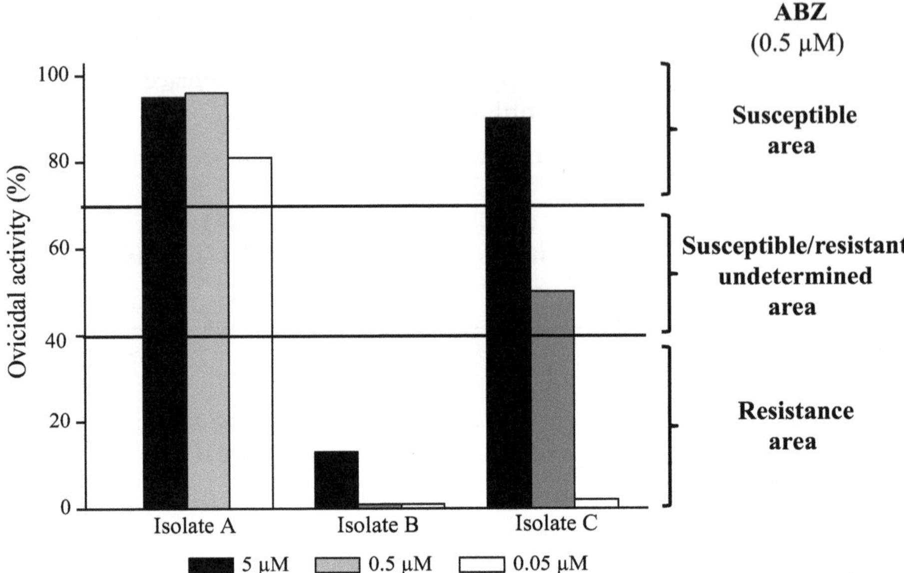

Fig. 1 The ovicidal activity (%) of albendazole (ABZ) on eggs from different *Fasciola hepatica* isolates (A, B, and C), in which: isolate A results ABZ-susceptible; isolate B ABZ-resistant and isolate C has an undetermined status of susceptibility/resistance to ABZ (susceptibility/resistance status defined according the ABZ ovicidal activity achieved at the "critical" concentration of 0.5 μM)

5. Before proceeding to assess the ABZ ovicidal activity on a specific *F. hepatica* isolate, make sure that egg development in the untreated eggs is ≥70% to avoid ambiguous/erroneous results.
6. The percentages of developed eggs should be reported as the arithmetic mean ± standard deviation (SD). The "ABZ ovicidal activity" should be expressed as percentage using the following formula:

$$\text{Ovicidal activity (\%)} = [(\%\text{eggs developed in control} \\ -\%\text{eggs developed after drug incubation}) \\ /\%\text{eggs developed in control}] \times 100$$

   Parametric (unpaired t test) or nonparametric (Mann–Whitney test) methods should be used for the statistical comparison of egg development in untreated and treated eggs (for the ABZ concentration of 0.5 μM). Statistical significance should be set at $P < 0.05$.
7. The "critical" ABZ concentration to estimate ovicidal activity is 0.5 μM. ABZ resistance is confirmed if egg development in untreated eggs is ≥70%, egg development did not significantly differ ($P > 0.05$) between ABZ-treated (0.5 μM concentration) and untreated eggs, and the ABZ ovicidal activity is ≤40%. The isolate will behave as susceptible if egg development in untreated eggs is ≥70%, egg development significantly differ

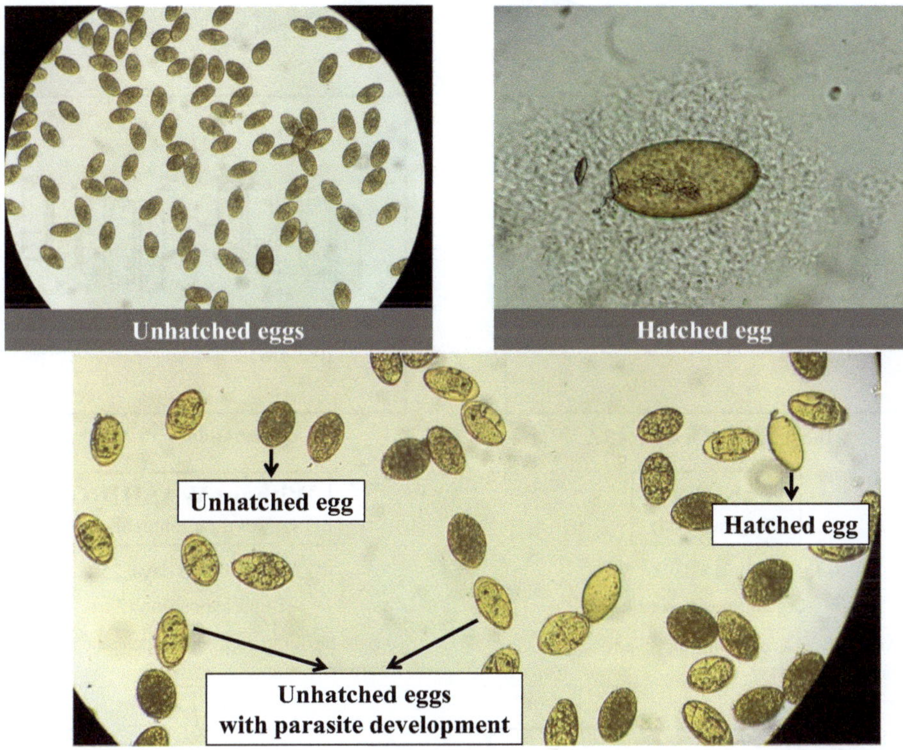

Fig. 2 *F. hepatica* eggs including unhatched eggs, hatched eggs, and unhatched eggs with parasite developed inside (miracidium stage), observed by an optical microscope

($P < 0.05$) between ABZ treated (0.5 μM concentration) and untreated eggs, and the ABZ ovicidal activity is ≥70%. With an egg development ≥70% (untreated eggs) and ABZ ovicidal activity between 40% and 70%, nonconclusive results about the susceptible/resistant status of the *F. hepatica* isolate under evaluation can be obtained (Fig. 1).

8. Testing ABZ activity at 5 and 0.05 μM is useful to predict the degree of resistance of the isolate. Eggs from highly ABZ-resistant isolates develop even after incubation at ABZ concentrations as high as 5 μM. Failures in egg development can be observed after ABZ incubation at concentrations as low as 0.5 μM in highly susceptible *F. hepatica* isolates.

4 Notes

1. *F. hepatica* eggs may also be recovered from faecal material of infected animals although, since ABZ may bind to debris, they should be as clean as possible. Eggs can be stored in tap water in centrifugal conical tubes, at 4 °C in the darkness for at least 2 months.

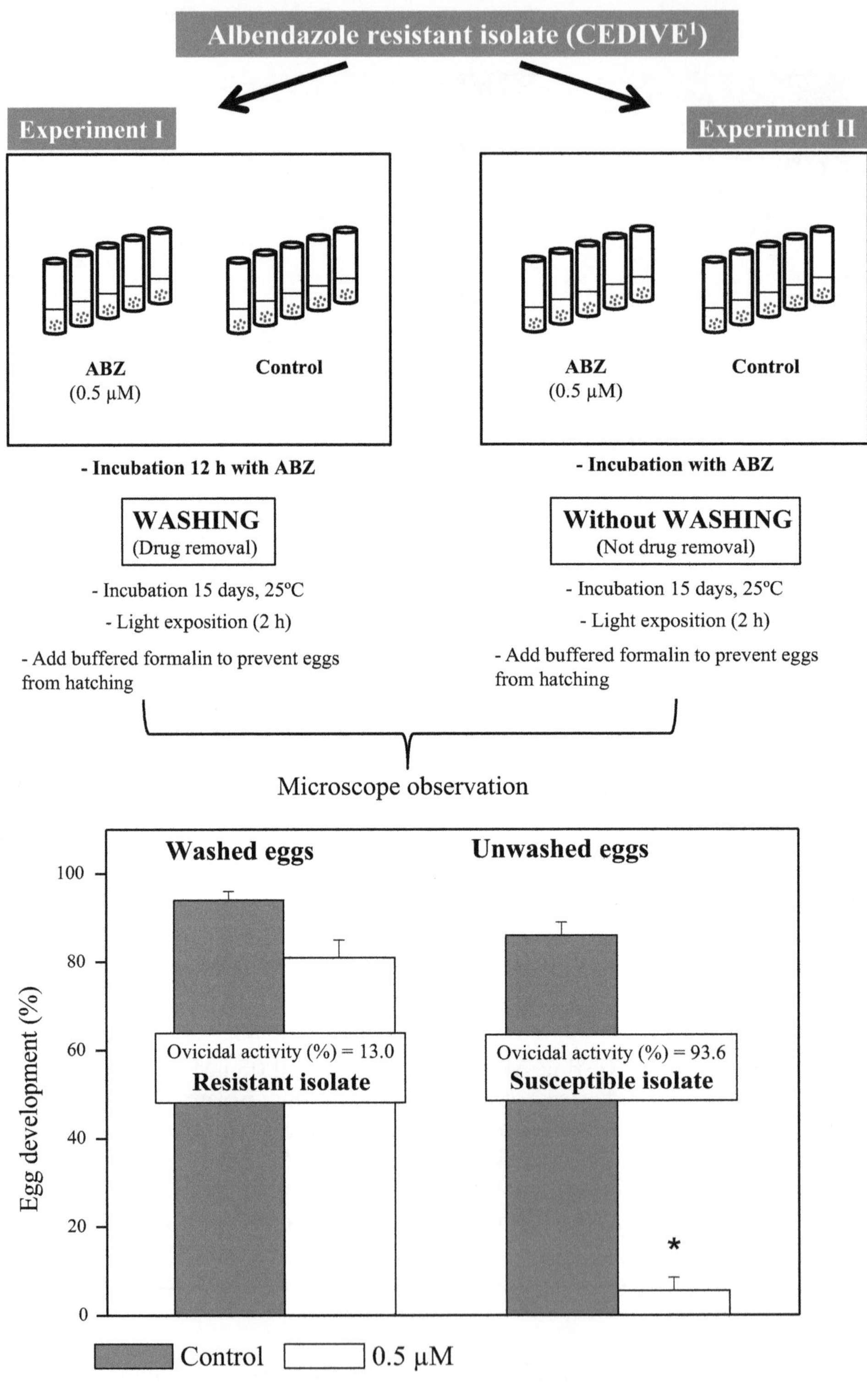

Fig. 3 Importance of the "egg washing step" to determine the susceptible/resistant status of a *Fasciola hepatica* isolate. Eggs obtained from an ABZ-resistant *F. hepatica* isolate [3, 10] were incubated with albendazole (ABZ, 0.5 μM) for 12 h (Experiment I) or 15 days (Experiment II). While drug removal in Experiment I allowed to confirm the resistant status of the isolate, the 15 days of eggs drug exposure (Experiment II) resulted in an apparent status of susceptibility (data not published)

2. The use of commercial formulations as a source of ABZ is not recommended. The excipients of the formulation could interfere with eggs development.
3. *F. hepatica* eggs are observed by 40x magnification in a standard microscope.
4. Since BZD-methylcarbamate water solubility increases at low pH values, ABZ dissolution (in both MEOH and DMSO) can be notably enhanced adding 20 μL of HCl (37%).
5. Under the described conditions, ABZ solutions (stock and working solutions) can be stable for 1 year. If DMSO is used as solvent, the working solutions should be kept a few minutes at room temperature to allow DMSO defrost.
6. It is useful to estimate an egg development "baseline" before starting the egg development test. The egg development "baseline" should not exceed 10% (Fig. 2).
7. The addition of the 10 μL working solution into the eggs suspension should be carefully performed. Gently shake the tube to assure drug solution–egg contact. If a low number of *F. hepatica* eggs are available (e.g., when eggs are collected from fecal material), the egg development test can be performed only at 0.5 μM but always including a set of eggs without drug. This is considered the "critical" concentration to estimate ABZ ovicidal activity and to assess the susceptible/resistant status of a *F. hepatica* isolate.
8. The incubation should be performed in an incubator to maintain the temperature stable during the whole test. If an incubator is not available, the incubation can be done in a warm place (~25 °C) in darkness. Darkness can be ensured by wrapping each Kahn tube with aluminum foil.
9. To wash the eggs after 12 h of incubation with the drug ("eggs washing step"), add 3 mL of tap water per tube to reach a final volume of 4 mL. Wait 10 min for the eggs to settle to the bottom and then remove 3 mL of water with a Pasteur pipette, avoiding dragging eggs by suctioning. Repeat this procedure three times. The use of glass Kahn tubes facilitates the washing procedure. This is an essential step, since an incomplete discard of drug or solvent can alter the results (Fig. 3). Try to do all procedures in as dark as possible environment.
10. The term "developed eggs" include hatched eggs and non-hatched eggs with the parasite developed inside (Fig. 2).

Acknowledgments

This study was funded by Agencia Nacional de Promoción Científica y Técnica (ANPCyT), Argentina.

References

1. McKellar Q, Scott E (1990) The benzimidazole anthelmintic agents: a review. J Vet Pharmacol Ther 13:223–247
2. Álvarez-Sánchez MA, Mainar-Jaime RC, Pérez-García J et al (2006) Resistance of Fasciola hepatica to triclabendazole and albendazole in sheep in Spain. Vet Rec 159:424–425
3. Sanabria R, Ceballos L, Moreno L et al (2013) Identification of a field isolate of Fasciola hepatica resistant to albendazole and susceptible to triclabendazole. Vet Parasitol 193:105–110
4. Novobilsky A, Amaya Solis N, Skarin M et al (2016) Assessment of flukicide efficacy against Fasciola hepatica in sheep in Sweden in the absence of a standardised test. Int J Parasitol Drugs Drug Resist 6:141–147
5. Wood IB, Amaral NK, Bairden K et al (1995) World Association for the Advancement of veterinary parasitology (W.A.A.V.P.) second edition of guidelines for evaluating the efficacy of anthelmintics in ruminants (bovine, ovine, caprine). Vet Parasitol 58:181–213
6. Fairweather I (2011) Reducing the future threat from (liver) fluke: realistic prospect or quixotic fantasy? Vet Parasitol 180:133–143
7. Lacey E (1990) Mode of action of benzimidazoles. Parasitol Today 6:112–115
8. Coles GC, Bauer C, Borgsteede F et al (1992) World Association for the Advancement of veterinary parasitology (W.A.A.V.P.) methods for the detection of anthelmintic resistance in nematodes of veterinary importance. Vet Parasitol 44:35–44
9. Alvarez L, Moreno G, Moreno L et al (2009) Comparative assessment of albendazole and triclabendazole ovicidal activity on Fasciola hepatica eggs. Vet Parasitol 164:211–216
10. Canevari J, Ceballos L, Sanabria R et al (2014) Testing albendazole resistance in Fasciola hepatica: validation of an egg hatch test with isolates from South America and the United Kingdom. J Helminthol 88:286–292
11. Robles-Pérez D, Martínez-Pérez JM, Rojo-Vázquez FA (2014) Development of an egg hatch assay for the detection of anthelmintic resistance to albendazole in Fasciola hepatica isolated from sheep. Vet Parasitol 203:217–221

Chapter 17

Drug Targets: Screening for Small Molecules that Inhibit *Fasciola hepatica* Enzymes

Florencia Ferraro, Mauricio A. Cabrera, Guzmán I. Álvarez, and Ileana Corvo

Abstract

The in vitro screening of small molecules for enzymatic inhibition provides an efficient means of finding new compounds for developing drug candidates. This strategy has the advantage of being rapid and inexpensive to perform. Enzymes are suitable targets for screening when simple methods to obtain them and measure their activities are available and there is evidence of their essential role in the parasite's life cycle. Here, we describe the screening of small molecules as inhibitors of two *Fasciola hepatica* enzyme targets (cathepsin L and triose phosphate isomerase), an initial step to find new potential compounds for drug development strategies.

Key words Drug targets, *Fasciola hepatica*, Enzyme screening, Small molecules, Drug discovery

1 Introduction

Since there is no available vaccine against *Fasciola hepatica* infections, chemotherapy is the only available control measure, with triclabendazole being the drug of choice for cattle and humans, as it is effective against both the juvenile and mature form of the parasite. Unfortunately, resistance to triclabendazole has been reported in several countries, demonstrating the urgent need to find new drugs and targets for drug development [1–3]. On the one hand, enzymes constitute attractive targets for drug discovery and, on the other, small molecules are excellent enzyme inhibitor candidates, with the ability to interact with enzyme pockets to interfere with its activity [4]. In this sense, the use of a biochemical activity assay with the enzyme targets to screen for compounds capable of interfering with their activity is a first approach to identify enzyme's inhibitors and allows to test a big set of molecules and then select the active ones to carry on additional studies. Nowadays, beyond the discovery of new drug candidates, the screening of

Martin Cancela and Gabriela Maggioli (eds.), *Fasciola hepatica: Methods and Protocols*, Methods in Molecular Biology, vol. 2137, https://doi.org/10.1007/978-1-0716-0475-5_17,

compound libraries is a strategy for repositioning (or repurposing) existing drugs for the treatment of alternative diseases, an interesting approach that may save significant time and money during drug development [5].

Here, we describe how to assay two different *Fasciola hepatica* enzymes types as molecular targets for drug discovery. One of them, the cathepsin L family of proteases (EC 3.4.22.15) is abundantly expressed at all parasite life stages [6, 7]. Cathepsins found in newly excysted juveniles can degrade collagen allowing the infective larvae to penetrate the gut wall tissue [8, 9]. In the adult parasites, cathepsins L are essential for nutrient acquisition by digesting blood proteins and for immune evasion, as they are able to cleave and inactivate immunoglobulins, thus modulating immune responses to generate a parasite friendly environment [10–12]. We also employ another enzyme target, triose phosphate isomerase (TPI, EC 5.3.1.1) which catalyzes the reversible isomerization of dihydroxyacetone phosphate and glyceraldehyde 3-phosphate, to avoid the dihydroxyacetone phosphate entering a metabolic "dead end." This is a key step in glycolysis that follows the splitting of fructose 1,6-bisphosphate into glyceraldehyde 3-phosphate and dihydroxyacetone phosphate, and allows all 6 carbon atoms of fructose to be further processed to pyruvate to generate ATP and NADH [13]. TPIs, found in all life kingdoms are active only in a homodimeric form and despite having high sequence and structural similarity, are different at the dimeric interface [14]. The design of selective inhibitors has been accomplished, for example, with trypanosomatids and *Giardia lamblia* TPI [14–16], being considered an interesting target for drug discovery.

The procedures described here can be used to assay cathepsins and TPI activity when a high-throughput automated platform is not available. It may also be useful as a guide to design activity screenings to find inhibitors of other *F. hepatica* enzyme targets.

2 Materials

2.1 Cathepsin L Inhibitory Activity Screening

1. 96-well plate fluorescence reader.
2. Black 96-well plates with flat bottom.
3. Multichannel pipette.
4. *Fh*CL1: 50 μM active *F. hepatica* cathepsin L1 proteolytic enzymes. Store at −20 °C (*see* **Note 1**).
5. *Fh*CL3: 50 μM active *F. hepatica* cathepsin L3 proteolytic enzymes. Store at −20 °C (*see* **Note 1**).
6. 100 mM sodium phosphate pH 6. Prepare 100 mL of a 1 M solution of monobasic sodium phosphate (H_2NaPO_4) and prepare 20 mL of a 1 M solution of dibasic sodium phosphate

(HNa_2PO_4). Prepare 100 mL of a 1 M stock solution of sodium phosphate buffer pH 6 by mixing 92 mL of 1 M monobasic sodium phosphate and 8 mL of 1 M dibasic sodium phosphate. The pH should be around 6; adjust if needed with the remaining solutions. Store at room temperature.

7. 100 mM EDTA solution. Store at room temperature (*see* **Note 2**)
8. 1 M DTT solution in water. Store at −20 °C.
9. Buffer to measure cathepsin L activity (CL Buffer): 100 mM sodium phosphate pH 6, 1 mM EDTA, 1 mM DTT. To prepare a 100 mL of activity buffer dilute 10 mL of 1 M sodium phosphate pH 6 in 89 mL of ultrapure water and add 1 mL of the EDTA stock solution. This buffer can be stored at room temperature. Before performing an assay, DTT should be added to the amount of buffer to be used; do not store buffer with DTT.
10. Fluorescent peptide substrates: for measuring cathepsin L proteolytic activity peptide fluorescent substrates are used. The substrate for measuring *Fh*CL1 activity is Z-Val-Leu-Lys-AMC and for *Fh*CL3 activity is Tosyl-Gly-Pro-Arg-AMC (*see* **Note 3**). Substrates are easily dissolved in 70% formamide to get a 10 mM stock solution. Do not weigh substrate; dissolve directly in the purchase container.

2.2 Triose Phosphate Isomerase Inhibitory Activity Screening

1. 96-well plate absorbance reader.
2. UV 96-well plates.
3. Multichannel pipette.
4. *Fasciola hepatica* triose phosphate isomerase (*Fh*TPI) (*see* **Note 4**).
5. Buffer to measure *Fh*TPI activity (TE buffer): 100 mM triethanolamine and 10 mM EDTA pH 7.4 in distilled water. To prepare 1 L you need 13.3 mL of triethanolamine and 2.92 g of EDTA, adjust pH to 7.4 with HCl (5 N).
6. DL-Glyceraldehyde-3-phosphate diethyl acetal barium salt: this product includes a specially washed Dowex 50 X 4-200R resin for the preparation of the substrate DL-glyceraldehyde-3-phosphate. To prepare the substrate wash 1.5 g (wet weight) of the Dowex-50 X 4-200R resin twice with distilled water by decanting, leave the resin as dry as possible. Add 6.0 mL of distilled water to the dried resin in a Pyrex test tube. Then, add 100 mg of DL-glyceraldehyde-3-phosphate diethyl acetal barium salt to the test tube and place it into a boiling water bath for approximately 3 min shaking intermittently (*see* **Note 5**). Quickly chill the solution by transferring the test tube to an ice bath. Centrifuge to decant the solution and save the supernatant or

recover the supernatant by filtering through a 0.45 μm syringe filter (a). Resuspend the resin in approximately 2 mL of water and recover the supernatant again (b). Combine the supernatants (a) and (b), they will contain approximately 25 mM of DL-glyceraldehyde-3-phosphate (the enzymatically active D-isomer). Store in 0.5 mL aliquots at −80 °C up to 6 months.

7. 10 mg/mL β-Nicotinamide adenine dinucleotide (NADH) stock: prepare 200 μL freshly before use (use NADH reduced disodium salt hydrate, ≥97% by HPLC), dissolve in distilled water. It cannot be stored.
8. 100–300 units/mg α-Glycerophosphate Dehydrogenase (α-GDH) stock: A 1:200 dilution is prepared mixing 10 μL of α-GDH from rabbit muscle Type I ammonium sulfate suspension with 1990 μL of TE buffer immediately before use. Use this dilution to perform the reaction mix. It cannot be stored.
9. Reaction mix: 1 mM glyceraldehyde 3-phosphate, 0.2 mM NADH, 0.9 units of α-glycerol phosphate dehydrogenase in TE buffer. For a 96-well plate, mix 5 mL glyceraldehyde 3-phosphate, 200 μL NADH, 2 mL of α-GDH 1:200 dilution and 12.6 mL of TE activity buffer. Keep protected from light; use immediately.

2.3 Compounds to Be Tested

1. Analytical scale.
2. Micro tubes 1.5 mL.
3. Vortex.
4. Dimethyl sulfoxide (DMSO).
5. Compounds: to screen for cathepsin L inhibition we employed synthetic flavonoids derivatives and for TPI inhibition we evaluated thiadiazines. All compounds were taken from our *in house* chemical library (*see* **Note 6**). The structure of some of the inhibitory molecules that were tested is presented in Fig. 1.

3 Methods

3.1 Compound Stock Preparation

1. Prepare an Excel file with the compounds list in the first column and their molecular weight in the second.
2. Weigh around 1–5 mg of each compound directly in a 1.5 mL micro tube to prepare around 1 mL of stock solution (*see* **Notes 7** and **8**). Enter the weighed mass of each compound into the third column of the excel file.
3. The fourth column is to calculate the volume of DMSO to dissolve the compound to get a 10 mM stock solution. Here, enter the formula = Mass (g)/(Molecular weight

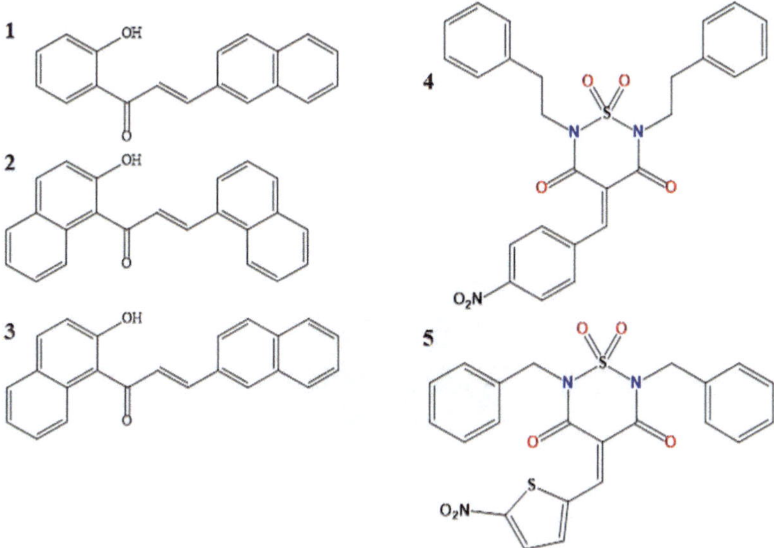

Fig. 1 Structure of three *F. hepatica* cathepsin L and two TPI inhibitors found by in vitro screening against the enzymes: flavonoids (1, 2, 3) and thiadiazines (4, 5) derivatives

(g/mol) × Molarity (M); that is = third column value/(second column value × 0.01).

4. Pipet the calculated volume of DMSO into each microtube and mix by inverting the tube several times and then vortexing until the compound is completely dissolved (*see* **Note 9**).

3.2 Cathepsin L Inhibitory Activity Screening

The cathepsin activity is measured as fluorescence units of AMC (RFU) released from substrate per unit of time (s) (Fig. 2). Plate layout can be designed to screen up to 44 compounds per 96-well plate in duplicate, leaving two wells for blanks and two wells for the positive controls (*see* **Note 10**). The total reaction volume is 200 μL per well. It is composed of 100 μL of enzyme preparation, 20 μL of the compound to be tested and 80 μL of substrate.

1. To set the reaction conditions, turn on the fluorescent reader and set up the parameters to do a continuous (kinetic) measurement at an excitation wavelength of 345 nm and emission wavelength of 440 nm for 100 reads. Set the excitation bandwidth (nm) in 12 and the optics to measure from top.
2. To prepare enzyme dilutions thaw an aliquot of *Fh*CL1 and/or *Fh*CL3 recombinant stock enzymes in an ice bath. Prepare a dilution considering a final enzyme concentration in the well of around 5 nM for *Fh*CL1 and 20 nM for *Fh*CL3 (*see* **Note 11**). Calculate the volume of diluted enzyme to prepare by multiplying the number of wells to be used for 100 μL, then dilute the enzyme in activity buffer.

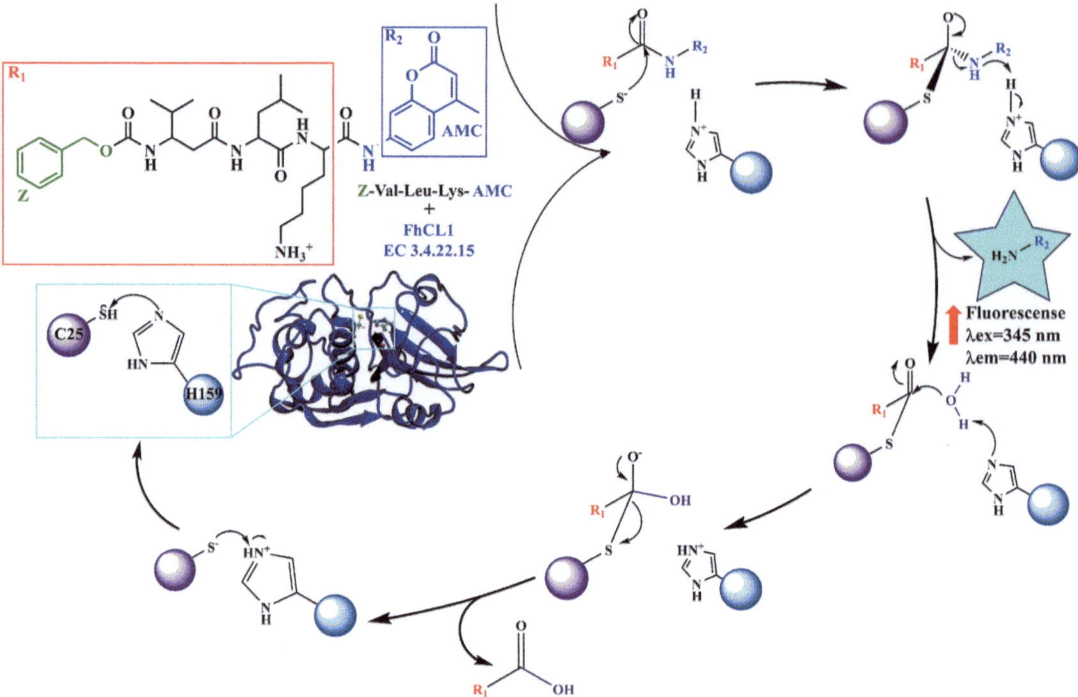

Fig. 2 Scheme of cathepsin L cleavage of a fluorogenic peptide substrate (EC 3.4.22.15). The structure of the Z-VLK-AMC substrate is shown on the left. FhCL1 enzyme is taken as an example and its structure is depicted in new cartoons (a zoom of the catalytic dyad is shown). A simplified mechanism of catalysis is represented based on [19]. The fluorescent product AMC is highlighted in a sky-blue star. The rate of AMC release from the substrate is measured in a fluorescence reader with an excitation wavelength of 345 nm and an emission wavelength of 440 nm. Please note that the compounds for screening are not included in this scheme

3. For each compound to be tested prepare 200 μL of a 100 μM solution by mixing 2 μL of the 10 mM stock with 198 μL of CL buffer (*see* **Note 12**).
4. Prepare an intermediate substrate dilution (8 mL per plate) diluting the 10 mM substrate stock solution in CL buffer to get a 50 μM solution (to prepare 8 mL mix 40 μL of the 10 mM substrate stock with 7.96 mL of CL buffer).
5. For plate setup, add 20 μL of the 100 μM compound dilution to each well, test compounds in duplicate, leave four wells free for two blanks and two positive controls. Add 20 μL of CL buffer to the positive control wells. Add 120 μL of CL buffer to the blank wells.
6. Then add 100 μL of the enzyme preparation to each well (except for the blank wells). The blank wells allow to check for noncatalyzed cleavage of the substrate. Incubate the plate for 30 min at room temperature (*see* **Note 13**).
7. To measure enzyme activity initiate the reaction by adding 80 μL of the intermediate substrate dilution to each well and

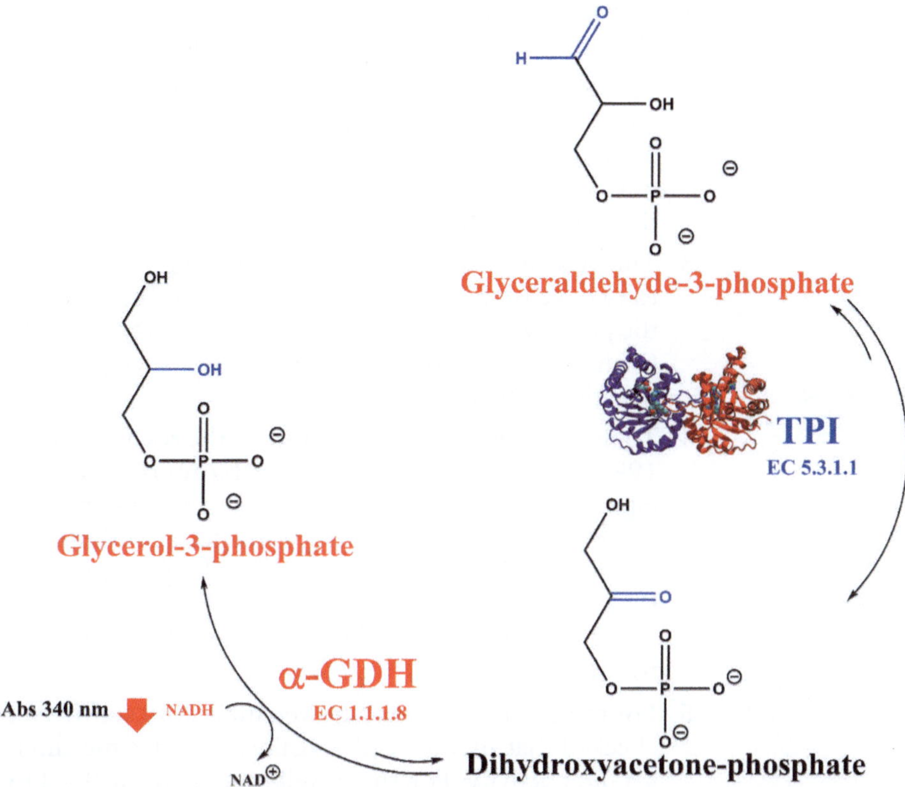

Fig. 3 Reaction scheme of the assay to measure triose phosphate-isomerase activity (EC 5.3.1.1). The reagents used to prepare the reaction mix are shown in red and the TPI enzyme is colored in blue. Dihydroxyacetone-phosphate and NAD+, the products of the reactions catalyzed by TPI and α-GDH, respectively, are colored in black. TPI activity is determined by the consumption of NADH that leads to a decrease in the photometric measure at 340 nm. Please note that the compounds for screening are not included in this scheme

start the measurement (*see* **Note 14**). Enzyme rate activity is expressed as RFU/s (relative fluorescence units of AMC released per second).

8. The percentage of enzyme inhibition is calculated as: % inh = 100 − (v_i/v_o) × 100, where v_i and v_o correspond to the initial rate of AMC fluorescence increase (RFU/s) with and without inhibitor, respectively.

3.3 Triose Phosphate Isomerase Inhibitory Activity Screening

TPI activity is determined in a reaction coupled to α-GDH, measuring the consumption of NADH when the α-GDH enzyme converts the dihydroxyacetone- phosphate (produced by TPI) into glycerol 3-phosphate (Fig. 3). The inhibition assay is performed in a 96-well microplate in a plate-reader spectrophotometer. TPI enzymatic activity is determined following the conversion of glyceraldehyde 3-phosphate into dihydroxyacetone phosphate by measuring the decrease in absorbance at 340 nm at 25 °C.

1. Prepare 10 mL of the *Fh*TPI enzyme dilution at 5 μg/mL in TE buffer with 10% of DMSO (*see* **Note 15**).
2. In a 96-well plate add 1 μL per well of each of the 10 mM compounds stock to be tested. Add 199 μL of the enzyme dilution and incubate at 37 °C for 1 h (the compound concentration will be 50 μM). Incubate the enzyme dilution alone in the same conditions.
3. Then transfer 2 μL of each enzyme–compound incubation mix to two wells of a UV 96-well plate (to perform duplicates). To the positive controls transfer 2 μL of the enzyme dilution alone (include two control wells) and leave two empty wells for the negative controls.
4. To measure enzyme activity, initiate the reaction by addition of 198 μL of the reaction mix (*see* **Note 16**). A multichannel pipette is needed (*see* **Note 14**). Enzyme rate activity is expressed as the decrease in absorbance at 340 nm per second, measured during 10 min at 25 °C. The percentage of enzyme inhibition is calculated as: % inh $= 100 - (v_i/v_o) \times 100$, where v_i and v_o correspond to the initial rate with and without inhibitor, respectively.
5. For the compounds that showed inhibitory activity, it has to be checked that they are active against TPI enzyme and not affect α-GDH activity. For these wells add 2 μL of the TPI control enzyme incubated alone and measure the TPI activity again. If the activity is recovered it can be assumed that the compound is inhibiting TPI activity (*see* **Note 17**).

4 Notes

1. The production of the *F. hepatica* recombinant active cathepsins L is described in [9]. Recombinant enzymes are kept at −20 °C for long-term storage. It is recommended that enzymes are fractionated in small aliquots for long term storage to avoid repeated freeze–thaw cycles. Nonetheless, *Fasciola hepatica* cathepsin Ls are robust enzymes that preserve most of their activity even after several freeze/thaw cycles and long storage times at −20 °C (several years).
2. The EDTA will not dissolve until the pH of the solution is 8. While stirring add a few drops of a 1 M NaOH solution until it become clear, that is when the EDTA is dissolved.
3. For measuring *Fh*CL1 activity other fluorescent peptide substrates might be used, containing the amino acid leucine in P2 and Arg or Lys in P1 (ex. Leu-Arg-AMC). Peptides with Phe in P2 are cut by *Fh*CL1 but with less efficiency. To measure *Fh*CL3 activity use peptides with Gly in P3, Pro in P2 and Arg or Lys in P1.

4. Recombinant *Fh*TPI production in *E. coli* is described in [13].
5. The commercial presentation of glyceraldehyde 3-phosphate is a DL-Glyceraldehyde 3-phosphate diethyl acetal barium salt mixture. It requires a long preparation protocol and it is very expensive. When dissolving the salt in boiling distilled water for hydrolysis, take care to agitate vigorously as it has low solubility in water. When you can no longer see the white powder in the solution, the hydrolysis is complete. Stop the reaction now to get the highest recovery yields.
6. The compounds employed to test for enzyme inhibition [17] can be provided by mail if requested. Their synthetic procedures are described in the references [16, 18]. With the protocol described here any compounds set other than those used in this work can be tested for the inhibition of these *F. hepatica* enzymes.
7. Compounds should be weight and dissolved in DMSO. We recommend to prepare around 1 mL of a 10 mM stock of each compound to be evaluated. It is advisable to use freshly prepared compounds, dissolve them the first day of the screening and try to use them within a week. Dissolved compounds can be stored at −20 °C but they might decompose with increasing storage times (depending on chemical stability of particular chemotypes). It is recommended to perform a spectrum from 300 to 450 nm wavelengths for each inhibitor to verify that none of the compounds absorb light in the measurement range.
8. It is advisable to weigh between 1 and 3 mg of compound directly in a 1.5 mL microtube, this would allow to prepare a volume that is adequate to microtubes capacity (less than 1.5 mL). Typically, between 150 and 500 g/mol is the molecular weight of druggable compounds.
9. Some compounds might not dissolve well in DMSO, in this case is advisable to vortex vigorously or incubate them in an ultrasound bath for at least 10 min. Sometimes it might take a long time to dissolve completely. If still does not dissolve a less concentrated stock might be prepared (e.g., 1 mM), but be careful when choosing the stock concentration, as the final DMSO concentration in the plate assay cannot exceed 1%, and optimally be below 0.5% DMSO.
10. Be sure to set the reading direction of the equipment so that the duplicates are read consecutively according to plate layout. This will contribute for better results reproducibility. If the number of compounds is not a limitation, perform in triplicate.
11. The *Fh*CL3 enzyme activity is lower than *Fh*CL1 that is why a higher enzyme concentration is needed to get an adequate slope of RFU/s to measure the inhibition of protease activity (increase the amount of enzyme if the slope is lower than 0.5).

12. The initial dose for the screening can vary. Often concentrations between 10 and 100 μM can be a good choice, usually higher concentrations are preferred when performing a random screening to increase the chance to identify any inhibitory molecule. Modify the intermediate compound dilution as necessary to get ten times the concentration to be used in the plate assay. To dilute the compound, we recommend to mix by pipetting and check if there is no insoluble material. If this occurs lower the concentration of the intermediate dilution.
13. It is recommended to incubate the enzyme and compounds for at least 30 min, to allow any slow-binding compounds to achieve inhibition.
14. The multichannel pipette is necessary so that the manipulation is as fast as possible and the start of the reaction in the first wells is not too late with respect to the last ones. If you take too long in this step and delay the measurement start, you will miss the initial rates. If a multichannel pipette is not available, test about 20 wells at a time.
15. When preparing the TPI enzyme working dilution (5 ng/μL) is recommended to minimize the number of intermediate enzyme dilutions to avoid losing enzyme activity. Diluted enzyme cannot be store, do not prepare in excess.
16. This protocol used to measure TPI inhibition allows to identify irreversible or tight-binding inhibitors, since fully reversible interactions would be lost when the compound is diluted 100 times to start the reaction measurement.
17. The TPI activity measurement relies on the activity of another enzyme, the α-GDH. Though the compound concentration at the start of the TPI enzymatic activity measurement is lowered 100 times, it is necessary to check that no relevant inhibition of α-GDH is preventing the TPI activity measurement, which could be mistaken with inhibition of TPI by a compound.

Acknowledgments

This work was supported by Universidad de la República and PEDECIBA, Uruguay and the Science and Innovation Fund, Uruguay—United Kingdom (British Embassy Montevideo, Grant SPF.2014.F001).

References

1. Brennan GP, Fairweather I, Trudgett A, Hoey EM, McCoy M, McConville M et al (2007) Understanding triclabendazole resistance. Exp Mol Pathol 82:104–109
2. Boray J, Crowfoot P, Strong M, Allison J, Schellenbaum M, von Orelli M et al (1983) Treatment of immature and mature Fasciola hepatica infections in sheep with triclabendazole. Vet Rec 113:315–317
3. Kelley JM, Elliott TP, Beddoe T, Anderson G, Skuce P, Spithill TW (2016) Current threat of Triclabendazole resistance in Fasciola hepatica. Trends Parasitol 32(6):458–469
4. Copeland R (2013) Evaluation of enzyme inhibitors in drug discovery. In: A guide for medicinal chemists and pharmacologists, 2nd edn. John Wiley & Sons, Inc., Hoboken, New Jersey
5. Baker NC, Ekins S, Williams AJ, Tropsha A (2018) A bibliometric review of drug repurposing. Drug Discov Today 23(3):661–672
6. Cancela M, Acosta D, Rinaldi G, Silva E, Durán R, Roche L et al (2008) A distinctive repertoire of cathepsins is expressed by juvenile invasive Fasciola hepatica. Biochimie 10:1461–1475
7. Robinson MW, Tort JF, Lowther J, Donnelly SM, Wong E, Xu W et al (2008) Proteomics and phylogenetic analysis of the cathepsin L protease family of the helminth pathogen Fasciola hepatica: expansion of a repertoire of virulence associated factors. Mol Cell Proteomics 6:1111–1123
8. McGonigle L, Mousley A, Marks NJ, Brennan GP, Dalton JP, Spithill TW et al (2008) The silencing of cysteine proteases in Fasciola hepatica newly excysted juveniles using RNA interference reduces gut penetration. Int J Parasitol 38(2):149–750
9. Corvo I, Cancela M, Cappetta M, Pi-Denis N, Tort JF, Roche L (2009) The major cathepsin L secreted by the invasive juvenile Fasciola hepatica prefers proline in the S_2 subsite and can cleave collagen. Mol Biochem Parasitol 167(1):41–47
10. Lowther J, Robinson MW, Donnelly SM, Xu W, Stack CM, Matthews JM, Dalton JP (2009) The importance of pH in regulating the function of the Fasciola hepatica cathepsin L1 cysteine protease. PLoS Negl Trop Dis 3 (1):e369
11. Donnelly S, O'Neill SM, Stack CM, Robinson MW, Turnbull L, Whitchurch C, Dalton JP (2010) Helminth cysteine proteases inhibit TRIF -dependent activation of macrophages via degradation of TLR3. J Biol Chem 285 (5):3383–3392
12. Robinson MW, Dalton JP, O'Brien BA, Donnelly S (2013) Fasciola hepatica: the therapeutic potential of a worm secretome. Int J Parasitol 43(3-4):283–291
13. Zinsser VL, Hoey EM, Trudgett A, Timson DJ (2013) Biochemical characterisation of triose phosphate isomerase from the liver fluke Fasciola hepatica. Biochimie 95:2182–2189
14. Olivares-Illana V, Perez-Montfort R, Lopez-Calahorra F, Costas M, Rodriguez-Romero A, Tuena de Gomez-Puyou M (2006) Structural differences in triosephosphate isomerase from different species and discovery of a multitrypanosomatid inhibitor. Biochemistry 45:2556–2560
15. Enriquez-Flores S, Rodriguez-Romero A, Hernandez-Alcantara G, De la Mora-De la Mora I, Gutierrez-Castrellon P et al (2008) Species-specific inhibition of Giardia lamblia triosephosphate isomerase by localized perturbation of the homodimer. Mol Biochem Parasitol 157:179–186
16. Alvarez G, Aguirre-López B, Varela J, Cabrera M, Merlino A, López GV et al (2010) Massive screening yields novel and selective triosephosphate isomerase dimer-interface-irreversible inhibitors with anti-trypanosomal activity. Eur J Med Chem 45:5767–5772
17. Ferraro F, Merlino A, Dell Oca N, Gil J, Tort JF, Gonzalez M et al (2016) Identification of Chalcones as *Fasciola hepatica* Cathepsin L inhibitors using a comprehensive experimental and computational approach. PLoS Negl Trop Dis 10(7):e0004834
18. Cabrera M, Simoens M, Falchi G, Lavaggi ML, Piro OE, Castellano EE et al (2007) Synthetic chalcones, flavanones, and flavones as antitumoral agents: biological evaluation and structure-activity relationships. Bioorg Med Chem 15(10):3356–3367
19. Rzychon M, Chmiel D, Stec-Niemczyk J (2004) Modes of inhibition of cysteine proteases. Acta Biochim Pol 51(4):861–873

Correction to: Fasciola hepatica: Methods and Protocols

Martin Cancela and Gabriela Maggioli

Correction to:
Martin Cancela and Gabriela Maggioli (eds.), *Fasciola hepatica: Methods and Protocols*, Methods in Molecular Biology, vol. 2137,
https://doi.org/10.1007/978-1-0716-0475-5

Unfortunately chapters 11 and 12 were published without including the name of Prof. Ana Espino. The author's name has been added now so that the author group reads as follows: Olgary Figueroa-Santiago and Ana Espino.

The updated online version of these chapters can be found at
https://doi.org/10.1007/978-1-0716-0475-5_11
https://doi.org/10.1007/978-1-0716-0475-5_12

Martin Cancela and Gabriela Maggioli (eds.), *Fasciola hepatica: Methods and Protocols*, Methods in Molecular Biology, vol. 2137,
https://doi.org/10.1007/978-1-0716-0475-5_18,

Index

Martin Cancela and Gabriela Maggioli (eds.), *Fasciola hepatica: Methods and Protocols*, Methods in Molecular Biology, vol. 2137, https://doi.org/10.1007/978-1-0716-0475-5,

I

K

L

M

N

O

P

R

S

T

MIX
Papier aus verantwortungsvollen Quellen
Paper from responsible sources
FSC® C105338

If you have any concerns about our products,
you can contact us on
ProductSafety@springernature.com

In case Publisher is established outside the EU,
the EU authorized representative is:
Springer Nature Customer Service Center GmbH
Europaplatz 3, 69115 Heidelberg, Germany

Printed by Libri Plureos GmbH
in Hamburg, Germany